SPECTRAL SENSITIZATION

THE FOCAL LIBRARY

Editorial Board
Prof. Dr. W. F. BERG,
General Editor and Chairman of the Editorial Board, Director, Photographisches Institut der E.T.H.
Zürich, Switzerland.

Dr. SUZANNE BERTUCAT,
Editor, "Science et Industries Photographiques",
Paris, France.

Prof. Dr. K. V. CHIBISOV,
Corresponding member of the Academy of Sciences of the U.S.S.R., Committee of Scientific Photography and Cinematography Department of Chemistry,
Moscow, U.S.S.R.

Prof. Dr. Dr.-ing h.c. J. EGGERT,
Professor Emeritus, Eidgenössische Technische Hochschule,
Zürich, Switzerland.

Prof. Dr. H. FRIESER,
Director, Institute of Scientific Photography, Technische Hochschule,
Munich, Germany.

Prof. Dr. A. HAUTOT,
Director, Physical Laboratories, University of Liège,
Liège, Belgium.

Dr. T. H. JAMES,
Editor, "Photographic Science and Engineering", Research Laboratory, Eastman Kodak Company,
Rochester, N.Y., U.S.A.

Dr. F. W. H. MUELLER,
Associate Director of Research and Development, General Aniline & Film Corporation,
New York, U.S.A.

E. W. H. SELWYN, B.Sc.,
Chief Physicist and Head of Library and Information Department,
Research Laboratory, Kodak Ltd.,
Harrow, England.

Prof. Dr. G. SEMERANO,
Professor of Physical Chemistry, University of Bologna,
Bologna, Italy.

Prof. Dr. M. TAMURA,
Department of Industrial Chemistry, Faculty of Engineering, University of Kyoto,
Kyoto, Japan.

Dr. C. WALLER,
Research Manager, Ilford Ltd.,
Ilford, Essex, England.

Prof. Dr. W. D. WRIGHT,
Professor of Applied Optics, Imperial College of Science and Technology,
London, England.

Assistant Editors
R. B. COLLINS,
Technical Director, Photographic Film Division, 3M Company,
St. Paul, Minnesota, U.S.A.

JOHN H. JACOBS,
Principal Research Physicist, Pell & Howell Research Centre,
Pasadena, California, U.S.A.

Executive Editor
R. J. COX, B.Sc., F.R.P.S.,
Editor of the Journal of Photographic Science,
London, England.

Publishers
THE FOCAL PRESS,
31 *Fitzroy Square, London, W.1, England*
and 20 *East 46 Street New York N.Y. 10017, U.S.A.*

SPECTRAL SENSITIZATION

Hans Meier
State Geochemical Research Institute,
Bamberg

THE FOCAL PRESS
LONDON and NEW YORK

© 1968 FOCAL PRESS LIMITED

SBN 240 50667 7

*No part of this book may be reproduced in any form
without written permission of the publishers*

Translated from the German manuscript by
Grace E. Lockie, B.Sc., A.R.P.S.

Printed and bound in Great Britain
at the Pitman Press, Bath

CONTENTS

PREFACE 11

I. THE SENSITOMETRY OF SPECTRALLY SENSITIZED PHOTOGRAPHIC EMULSION LAYERS 13

1.1 Sensitometric principles 14
   1.1.1 Basic photometric conceptions 14
   1.1.2 Special photometric problems in photography 15
      1.1.2.1 The construction of characteristic curves 15
         1.1.2.1.1 Sensitometers 15
         1.1.2.1.2 Densitometers 15
         1.1.2.1.3 Methods of constructing characteristic curves 16
      1.1.2.2 Integral and spectral sensitivity 16

1.2 The spectral properties of non-sensitized and sensitized photographic emulsion layers 18
   1.2.1 General considerations of the sensitivity S_i and S_λ of sensitized emulsion layers 18
   1.2.2 Spectral sensitivity distribution 19
      1.2.2.1 Methods of constructing S_λ/λ curves 19
      1.2.2.2 The S_λ/λ relation in sensitized and non-sensitized emulsion layers 22
   1.2.3 Monochromatic gamma values in relation to wavelenth. 24
   1.2.4 The Schwarzschild effect in relation to wavelength 25
   1.2.5 Numerical specification of the effect of sensitization 25
   1.2.6 The relative quantum yield 27
   1.2.7 The influence of dye-sensitization on the characteristic curve 28

II. SENSITIZING DYES 33

2.1 The relation between the constitution of a dye and the region of sensitization 33
   2.1.1 The position of maximum sensitization in the spectrum 33
   2.1.2 The displacements in absorption caused by adsorption 33
   2.1.3 The theory of the absorption of light by dyes 35
      2.1.3.1 The electron gas model 35
      2.1.3.2 The relation between light-absorption and constitution 40

2.2 The connection between the constitution of a dye and its sensitizing action 49
   2.2.1 The efficiency of sensitization 49
   2.2.2 General prerequisites for the suitability of dyes for use as sensitizers 51
   2.2.3 The effects produced by substitution 52
   2.2.4 The influence of the spatial structure of a dye on its photographic behaviour 57
   2.2.5 The effect of aggregation 60

2.3	The sensitizing dyes used in practice			61
	2.3.1	The fundamental type of polymethine dye		61
		2.3.1.1 Cationic or basic polymethine dyes		62
			2.3.1.1.1 Structure	62
			2.3.1.1.2 Nomenclature	64
			2.3.1.1.3 Synthesis	66
		2.3.1.2 Anionic or acid polymethine dyes		68
		2.3.1.3 Neutral polymethine dyes		69
	2.3.2	Polymethine dyes of higher order		70
		2.3.2.1 Rhodacyanines		70
		2.3.2.2 The neocyanine type of dye		71
		2.3.2.3 Holopolar cyanines		71
	2.3.3	The preparation of sensitized emulsions		73

III. THE FUNDAMENTALS OF SPECTRAL SENSITIZATION — 78

3.1	General laws governing photographic sensitization			78
	3.1.1	Factors concerned with constitution		78
	3.1.2	Physico-chemical laws		78
		3.1.2.1 Quantum yield		78
		3.1.2.2 The repeated reactivity of the dye molecules		79
		3.1.2.3 Spectral sensitization in relation to temperature		79
		3.1.2.4 Spectral sensitization in relation to the thickness of the layer of dye		80
		3.1.2.5 The arrangement of the dye molecules		81
		3.1.2.6 The adsorption of sensitizing dyes in the presence of gelatin		82
		3.1.2.7 Displacements in absorption		83
		3.1.2.8 The fluorescence of adsorbed sensitizers		83
		3.1.2.9 The participation of triplet states		84
		3.1.2.10 The influence of chemical sensitization		84
3.2	Special sensitization reactions			85
	3.2.1	Desensitization		85
	3.2.2	Supersensitization		89
	3.2.3	Antisensitization		91

IV. THE SPECTRAL SENSITIZATION OF PHOTOCONDUCTIVITY — 96

4.1	The photoconductivity of spectrally sensitized silver halides			96
	4.1.1	The mechanism of the formation of the latent image		96
	4.1.2	Characteristic features of the sensitized photoconductivity of silver halides		98
		4.1.2.1 Its relation to wavelength		98
		4.1.2.2 The influence of desensitizing dyes		98
		4.1.2.3 Supersensitization		100
		4.1.2.4 Other cases of agreement with photographic effects		100
4.2	The spectral sensitization of the photoconductivity of inorganic photo-semiconductors			101
	4.2.1	Sensitizable photosemiconductors		101
	4.2.2	The nature of sensitizing dyes		105

		4.2.3	Laws governing the spectral behaviour of photosemiconductors	106
			4.2.3.1 Sensitization yield	107
			4.2.3.2 Repeated sensitizing power	107
			4.2.3.3 The effect of temperature	107
			4.2.3.4 The adsorption of dyes	108
			4.2.3.5 Desensitization	108
			4.2.3.6 The influence of gases	108
			4.2.3.7 Supersensitization	109
			4.2.3.8 The quenching of fluorescence	109
		4.2.4	Photoinduced EPR signals	110
	4.3	The spectral sensitization of the photoconductivity of organic semiconductors		110

V. THE PHOTOCONDUCTIVITY OF SENSITIZING DYES — 116

	5.1	The importance of research work on the photoconductivity of dyes in relation to spectral sensitization		116
	5.2	The measurement of the photoconductivity of a dye		118
		5.2.1	Methods of measurement	118
		5.2.2	Preparation	118
	5.3	A summary of some of the laws governing the dark- and photo-conductivity of dyes		120
		5.3.1	Dark-conductivity	120
		5.3.2	Photoconductivity in relation to the spectrum	122
		5.3.3	Photoconductivity in relation to the voltage	123
		5.3.4	Photoconductivity in relation to the intensity of the irradiation	126
		5.3.5	The quantum yield	126
		5.3.6	The inertia of the growth and decay process	128
		5.3.7	Temperature effects	130
		5.3.8	n- and p-conduction in dyes	131
		5.3.9	The lifetime and the mobility of the charge carriers	135
	5.4	The mechanism of the conductivity of dyes		136
		5.4.1	The tunnel effect mechanism	137
		5.4.2	The band model hypothesis	137
		5.4.3	The exciton hypothesis	141

VI. MODERN VIEWS ON THE MECHANISM OF SPECTRAL SENSITIZATION — 145

	6.1	Principles of the theories			145
		6.1.1	The resonance theory		146
			6.1.1.1 Arguments in support of the resonance mechanism		146
				6.1.1.1.1 Defects	146
				6.1.1.1.2 Radiationless transfer	147
				6.1.1.1.3 Position of the energy levels	147
			6.1.1.2 Arguments against the resonance mechanism		148
				6.1.1.2.1 The influence of defects	148
				6.1.1.2.2 The problem of the radiationless transfer of energy	149
				6.1.1.2.3 The position of the energy levels	151
		6.1.2	The electronic theory of sensitization		152
		6.1.3	Comment on the question: is the reaction one of resonance or of electron transfer?		153

6.2 The electron transfer theory of spectral sensitization — 154
    6.2.1 The probability of the transfer of electrons between a dye and a semiconductor — 154
    6.2.2 The relation between the conductivity of sensitizing dyes and spectral sensitization — 156
        6.2.2.1 The dark conductivity of dyes and sensitization — 156
        6.2.2.2 The relation between layer thickness and sensitizing power — 157
        6.2.2.3 The relation between sensitizing power and temperature — 157
        6.2.2.4 The relation between sensitizing power and wavelength — 157
        6.2.2.5 Supersensitization — 158
            6.2.2.5.1 The connection between supersensitization and the photoconductance of a dye — 158
            6.2.2.5.2 Rules governing the doping effect — 159
            6.2.2.5.3 Interpretation of the effect of doping — 163
        6.2.2.6 The connection between photoconductance and desensitization — 165
    6.2.3 The question of the energies involved in spectral sensitization — 169
        6.2.3.1 Definitions — 169
        6.2.3.2 Previous results — 174

VII. THE BOUNDARY LAYER OR PN MECHANISM OF SPECTRAL SENSITIZATION — 179

7.1 Investigations using model arrangements of sensitized systems — 179
    7.1.1 The measuring arrangement — 179
    7.1.2 The results of the measurements — 181
        7.1.2.1 Order of magnitude of the photovoltaic effect — 181
        7.1.2.2 Factors governing the direction of the photocurrent and the photovoltage — 181
        7.1.2.3 Relation to intensity — 183
        7.1.2.4 Inertia — 183
        7.1.2.5 Spectral sensitivity — 184
        7.1.2.6 Current-voltage characteristics — 184
    7.1.3 Interpretation of the results of the measurements — 186
        7.1.3.1 General considerations — 186
        7.1.3.2 The pn-theory of the photovoltaic effect in model systems — 187
    7.1.4 A summary of the facts which are indicative of a pn-transfer in model systems — 193
        7.1.4.1 Dependence on direction — 193
        7.1.4.2 Dependence on intensity — 196
        7.1.4.3 Dependence on the spectrum — 196
        7.1.4.4 Current-voltage characteristics — 196
        7.1.4.5 The influence of doping — 196

7.2	Application of the results obtained with model arrangements of sensitized systems to the mechanism of spectral sensitization		198
	7.2.1	The pn-theory of spectral sensitization	198
	7.2.2	Experimental indications of the validity of the pn-theory of spectral sensitization	199
		7.2.2.1 Relation to the spectrum	199
		7.2.2.2 Relation to direction	199
		7.2.2.3 Desensitization	202
		7.2.2.4 The repeated reactivity of a dye	203
		7.2.2.5 Supersensitization	203
		7.2.2.6 Other effects	204
	7.2.3	The problem of sensitization with monomolecular layers of dye	205
		7.2.3.1 The photoconductivity of monomolecular layers of dye	205
		7.2.3.2 The formation of a space charge between a dye and a semiconductor	205
		7.2.3.3 Experimental indications	209
7.3	Concluding remarks on Chapters VI and VII		212

INDEX 215

Dedicated to
Prof. Dr. h.c. mult. John Eggert

PREFACE

The discovery of spectral sensitization by H. W. Vogel in 1873 may be regarded as a decisive stage in the development of present day photography. Without the sensitization effect a large part of modern photography would be inconceivable, since dyes are the sole means by which silver halide emulsions, which are only sensitive to blue and ultraviolet light, can acquire the ability to "see" in the same way as the human eye. Again it is only through the agency of sensitizing dyes that black-and-white materials are able to give a natural colour rendering of the brightness values of objects, and finally, colour photography would not be possible without the use of suitable sensitizers.

The importance of studies which are devoted to the spectral sensitization effect is not limited to the photographic sector. The problems involved in the sensitization reaction are of interest in many spheres—in photochemistry, physics, organic and physical chemistry, biology, etc.

A question of general interest is, for instance, the mechanism of this effect, because spectral sensitization reactions are frequently encountered. As a brief reminder we would cite the mechanism of vision in the eye, the biologically important photodynamic effect, the process of assimilation in plants, or electrophotographic reproduction processes. In every case these processes commence with a dye which absorbs light and thus initiates secondary reactions.

Moreover the experiments which were concerned with photographic sensitization often gave rise to suggestions and statements of general importance, and as an example we would mention the question of the relation between colour and constitution, or the problem of the association of dyes, which has been specially studied in the case of sensitizing dyes. We would also mention that the electronic semiconductivity of organic dyes which is of such great interest nowadays, was established for the first time and investigated thoroughly in experiments aimed at elucidating the mechanism of sensitization.

In the present monograph the sensitometric problems connected with spectral sensitization are considered first, and this is followed by a discussion of the dyes which can be used in spectral sensitization and their requisite constitutions and spatial configurations. A summary of the laws governing the spectral sensitization of the photographic process and the photoelectric effect is followed by a critical treatment of the various theories of the mechanism of sensitization. In the last chapter the possibility of further developing previous electron-transfer hypotheses which were obtained from measurements of the photovoltaic effects observed in model experiments with sensitized systems is discussed.

Here I should like to thank Professor Dr. W. F. Berg for suggesting the writing of this monograph, and for the trouble which he has taken in reading through the manuscript, and also for his valuable advice. My special thanks are also due to Professor Dr. W. Jaenicke for promoting my work, for discussions and for numerous hints and suggestions.

Furthermore I should like to thank my collaborators Dr. A. Haus (now in Bayer-Leverkusen) and Dr. W. Albrecht, who also prepared the figures, for their indefatigable assistance in a large number of the experiments, as well as Dr. W. Hecker for suggestions and discussions. I wish to convey my thanks to Agfa-Gevaert Photofabriken (in particular to Dr. O. Riester), Leverkusen and to the Farbwerke Hoechst for providing the dyes, and to the Fonds der Chemischen Industrie, Düsseldorf, for financial aid in carrying out the experimental work. In addition I should like to thank the Deutschen Forchungsgemeinschaft and IIT, Chicago (particularly Professor L. Grossweiner in the latter) for the support which enabled me to make those contacts which are so necessary to international specialists.

The good co-operation of the publisher and Miss Grace E. Lockie, B.Sc., has been a great help to me. I would especially like to offer my heartiest thanks to Miss Lockie for her excellent translation.

I feel a specially warm sense of gratitude to my late teacher, Professor W. Noddack, who many years ago led me to take up the study of spectral sensitization, and last, but not least, my sincere thanks are due to Professor J. Eggert and Professor G. M. Schwab for promoting and showing a lively interest in my work.

HANS MEIER

I. THE SENSITOMETRY OF SPECTRALLY SENSITIZED PHOTOGRAPHIC EMULSION LAYERS

Sensitizing dyes influence the properties of photographic materials in many ways. Firstly they displace the spectral sensitivity of silver halide emulsion layers into the long wavelength region of the spectrum, and secondly they alter a number of the characteristic properties of the emulsions.

A knowledge of these effects is, of course, of great importance for the use of sensitizing dyes in practice. In the following we shall therefore discuss briefly a few of the essential points in the sensitometric testing of spectrally sensitized emulsion layers. For a more detailed orientation of general sensitometric problems and of methods of testing, the reader is referred to the comprehensive monographs and reviews by *Gorokhovskii, Eggert, Levenburg, Tupper, et al.*[1–6]

TABLE 1

PHYSICAL PHOTOMETRIC SYSTEM OF UNITS

Term	Symbol	Derivation	Unit
Power (flux) of the radiation	W		watts or $\dfrac{\text{ergs}}{\text{sec}}$
Intensity of the radiation	J	$\dfrac{\text{radiation efficiency in direction } \vartheta}{\text{solid angle}}$	$\dfrac{\text{watts}}{\text{steradian}}$
Luminance (brightness)	B	$\dfrac{\text{intensity of illumination in direction } \vartheta}{\text{apparent emitting area}}$	$\dfrac{\text{watts}}{\text{steradian} \cdot \text{cm}^2}$
Luminosity	R	$\dfrac{\text{radiation efficiency}}{\text{emitting area}}$	$\dfrac{\text{watts}}{\text{cm}^2}$
and			
Intensity of illumination	I	$\dfrac{\text{incident radiation efficiency}}{\text{receiving area}}$	$\dfrac{\text{watts}}{\text{cm}^2}$
Luminous intensity, respectively		$\dfrac{\text{intensity of radiation } I\vartheta \text{ of the emitter}}{(\text{emitter-receiver distance})^2}$	
Quantity of illumination	E	intensity of the illumination $\times$ time	$\dfrac{\text{watts} \cdot \text{sec}}{\text{cm}^2}$ and $\dfrac{\text{ergs}}{\text{cm}^2}$, respectively

1.1 Sensitometric Principles

1.1.1 BASIC PHOTOMETRIC CONCEPTIONS. Sensitometric judgment of the quality of photographic emulsions requires accurate data of the intensity of the light which is incident on the emulsions. Integral sensitometry is concerned with the determination of the photographic influence of non-dispersed daylight, the intensity of which is evaluated from its effect on our visual response to light. In spectral sensitometry, on the other hand, the action of monochromatic light on a photographic emulsion layer has to be measured.

Since the eye is incapable of making absolute measurements, the radiation quantities in the visual photometric system used in integral sensitometry[7,8] are derived from a special fundamental quantity, the new candela. On the other hand, in the physical photometric system, the monochromatic radiation is given directly by the effect which it produces, and therefore the radiation quantities in the visual and physical systems are different—see Tables 1 and 2. For the methods of measurement see *Bauer et al.*[7,9,9a]

TABLE 2

VISUAL PHOTOMETRIC SYSTEM OF UNITS

Term	Symbol	Derivation	Unit	Special Notation
Luminous flux or Luminous efficiency	W	luminous intensity × solid angle	candela . steradian	lumen
Intensity of light or Luminous intensity	J	fundamental quantity	candela	$\dfrac{\text{lumen}}{\text{steradian}}$
Luminance (brightness)	B	$\dfrac{\text{luminous intensity}}{\text{apparent emitting area}}$	$\dfrac{\text{candela}}{\text{cm}^2}$	stilb
Luminosity	R	$\dfrac{\text{luminous flux}}{\text{area}}$	$\dfrac{\text{lumens}}{\text{cm}^2}$	phot
Intensity of illumination	I	$\dfrac{\text{luminous flux}}{\text{receiving area}}$	$\dfrac{\text{candela . steradian}}{\text{m}^2}$	lux
		$\dfrac{\text{luminous intensity}}{(\text{receiver-emitter distance})^2}$	$\dfrac{\text{candela}}{\text{m}^2}$	
Exposure or quantity of light	E	intensity of the illumination × time	$\dfrac{\text{candela . sec}}{\text{m}^2}$ $\dfrac{\text{lumens . sec}}{\text{m}^2}$	lux second

1.1.2 SPECIAL PHOTOMETRIC PROBLEMS IN PHOTOGRAPHY.

1.1.2.1 *The Construction of Characteristic Curves.* In order to measure and evaluate characteristic curves, which were first defined by *Hurter* and *Driffield*, and which may be regarded as the basis of the sensitometric testing of photographic emulsion layers, it is necessary:

1. To irradiate the layer with graduated radiation energies or quantities of light, by means of sensitometers.[10]
2. To measure the density of a photographic image which has been developed under standard conditions, by means of densitometers.

1.1.2.1.1 Sensitometers. To produce an image of illumination corresponding to mean daylight, which is the purpose of integral sensitometry, special lamp-filter combinations are employed.[11,12] A suitable combination has proved to be, for example, a tungsten lamp burning at a colour temperature of 2850°K and a glass filter (Corning 5900). In order, however, to test the spectral behaviour of a photographic emulsion layer, the exposure is carried out using the entire continuous spectrum of a spectrograph[13] or monochromatic light from regions of various wavelengths.[14-16]

The chief method of graduating the irradiation energy is by the use of intensity sensitometers in which the intensity of the radiation is varied by means of continuous or step wedges. Since the time of exposure is constant, and because the intensity of radiation behind the individual steps in the wedge is graduated logarithmically, the exposure E is given directly on a logarithmic scale. The absorption of the available wedges is largely independent of wavelength, so that spectrally dispersed radiation can also be varied in steps.

1.1.2.1.2 Densitometers. The blackening or optical density is governed by the following relation:

$$D = \log \frac{W_0}{W_D} \qquad (1)$$

Thus, in theory, D can be determined by comparing measurements of the radiation efficiency of the test lamp made with (W_D) and without (W_0) an absorbing layer. With simple electric densitometers the density can be read off directly on a logarithmically calibrated galvanometer, whereas in visual subjective photometry the intensities of radiation of two part-beams are equalized by means of a continuous grey wedge calibrated in density units. Densitometers which record automatically the displacements of the comparison wedges are on the market.[2-4,17]

In contrast to photoelectric indicating instruments which are placed at a distance and which only record the radiation transmitted, cells which are located immediately behind the photographic emulsion layer measure in addition a portion of the radiation which is scattered by diffusion by the silver grains. The density D'' which is obtained in the first case is therefore greater than the diffuse density $D^\#$ which is usually given in practice, the Callier coefficient or Callier Q factor, which is defined by the relation

$$Q = \frac{D''}{D^\#} (\approx 1\cdot 5) \qquad (2)$$

being dependent on the size of the grains.[19]

In many cases the density is also ascertained by measuring the reflected portion of the incident test beam.[20, 21]

1.1.2.1.3 *Methods of Constructing Characteristic Curves.* The drawing of characteristic curves is facilitated by the use of densographs. With these

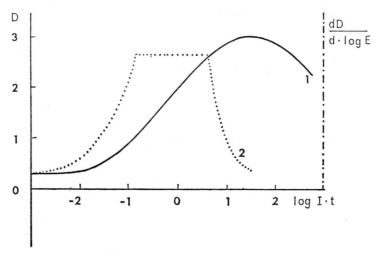

Fig. 1. Diagram of the characteristic curves $D = f(\log E)$. I. General shape. II. Derivation of the characteristic curve: $dD/d \log E = f(\log E)$.

instruments the densities of emulsion layers exposed under a grey wedge are compared with those of an identical wedge, and when the material is photometered, are plotted directly on a $D/\log E$ system of co-ordinates. This can also be done automatically.[22, 23] Mention should also be made of the crossed wedge method[24] which dispenses with density measurements.[24, 25] In the latter the test-emulsion is first exposed behind a quadratic grey wedge and the resulting negative is placed, rotated through 90°, on the wedge. This results in lines of equal density running parallel to one another, and, by copying the crossed wedges, an image of the characteristic curve is obtained.

1.1.2.2 *Integral and Spectral Sensitivity.* All the data which are necessary for describing the characteristics of photographic materials, namely the gamma,[26, 27] gradation, exposure range, fog[28] and sensitivity, can be obtained from the characteristic curves—cf. the diagram in Fig. 1. Here the determination of the requisite quantities for specifying the integral and spectral sensitivities of the materials is of particular importance.

In earlier sensitivity systems the energy of the threshold value (Scheiner degrees), and inertia (H. and D. degrees) or the exposure required to produce a density of $D = 0{\cdot}1$ above the fog (DIN system), to quote examples, were taken as a measure of the integral sensitivity S_i.[29] Since the choice of these speed indices presented a number of problems,[30] the American ASA and the British BSI systems stipulated special developing conditions and established a statistically derived gradation for the speed index.[31,32] Modern sensitivity systems (American standard PH2.5—1960, ISO and DIN 4512) analogously defined the index in terms of the minimum gradation which has been determined

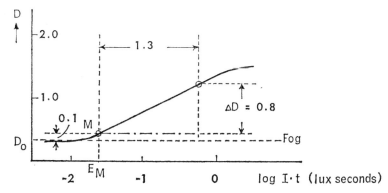

Fig. 2. Diagrammatic representation of the conditions for determining the speed number S (according to ASA, DIN 4512). (Definitions, see text.)

with great accuracy in a large number of experiments.[33] This gradation, moreover, coincides with a density of 0·1 units above the fog on the characteristic curve if certain conditions are observed and the material is developed to an average gamma value (corresponding to Fig. 2). According to DIN 4512 the following equation is specially valid for integral sensitivity

$$S_i = 10 \cdot \log \frac{1}{E_M} \qquad (3)$$

where E_M is the exposure value in lux-seconds corresponding to the minimum gradation.

The definition of a universally applicable spectral sensitivity S_λ for describing the characteristics of sensitized materials presents some difficulty. The monochromatic characteristic curves $D_\lambda = f(\log E_\lambda)$[34] of individual photographic materials often differ from one another in shape, so that in contrast to the integral speed index, a universal spectral sensitivity index cannot be specified in terms of a specific gradation or a specific density. The index is therefore established arbitrarily, and is namely the density which is situated in the centre of the characteristic curve after a certain time of development, i.e. at $D^\# = 1$ above the fog D_0 (see Fig. 3):

$$S_\lambda^* = \left(\frac{1}{E_\lambda}\right)_{D^\#=D_0+1} \quad [\text{cm}^2/\text{erg}] \qquad (4)$$

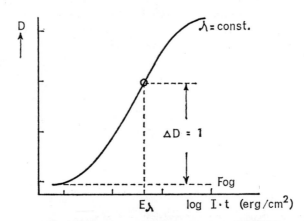

Fig. 3. Method of determining spectral sensitivity.

1.2 *The Spectral Properties of Non-sensitized and Sensitized Photographic Emulsion Layers*

1.2.1 GENERAL CONSIDERATIONS OF THE SENSITIVITY S_i AND S_λ OF SENSITIZED EMULSION LAYERS. The spectral sensitivities S_λ of sensitized photographic emulsions often differ appreciably from one another in different regions of the spectrum. This difference is undoubtedly troublesome in many cases and can only be dealt with by inserting filters in order to reproduce the energy of a fairly large spectral region correctly at all wavelengths. It is, however, of particular importance first to match the relative spectral sensitivity range of the emulsions as closely as possible to the sensitivity curve of the eye by the choice of suitable sensitizing dyes.

The following holds for the integral sensitivity S_i and the spectral sensitivity S_λ of a sensitized emulsion:

1. The integral sensitivity S_i can be increased considerably by sensitization. Thus in certain circumstances a low-speed emulsion can be converted into a high-speed emulsion layer which gives accurate reproduction of the brightness values.

2. The sensitivity of the sensitized regions, and at the same time, the integral sensitivity S_i of emulsion layers which have been spectrally sensitized with

reference to non-dispersed daylight can be increased by a multiple factor by the use of so-called supersensitizers.[35–38] The position of the spectral sensitivity curve remains unchanged in these conditions, since the compounds which absorb mostly in the ultra-violet do not themselves exert any spectral sensitizing action.

3. Small quantities of certain dyes of irregular constitution, the so-called antisensitizers, can lower the spectral sensitivity S_λ produced by a sensitizer. According to the STERN-VOLMER equation,[39] which holds for the quenching of fluorescence by impurities, the spectral sensitivity decreases with increase in the concentration of the antisensitizer[40]; this is expressed by an equation analagous to that for the quenching of fluorescence by impurities, namely:

$$\frac{S_\lambda}{S_{\lambda[A]}} = 1 + k \cdot C_A \qquad (5)$$

The quantities S_λ, $S_{\lambda[A]}$ in the equation denote the spectral sensitivities in the region of sensitization without and with an antisensitizer, C_A is the concentration of the antisensitizer, and k is a constant which combines the duration of the state of activation of the sensitizer and the rate constant of the transfer of energy.

4. Unlike supersensitizers and antisensitizers which are only effective in a sensitized spectral region, desensitizers diminish the spectral sensitivity of silver halide which lies beyond the region of the absorption of the dye.[41]

5. Increasing the concentration of the sensitizing dye often produces a desensitizing effect in the region of the natural absorption of silver halide,[42–45] and as a result, when the concentration of the sensitizer is raised, the integral sensitivity S_i first increases to a maximum value and then decreases steadily.[44–46]

1.2.2 SPECTRAL SENSITIVITY DISTRIBUTION.

1.2.2.1 *Methods of Constructing S_λ/λ Curves.* In theory, spectral sensitivity curves can be constructed from the spectral sensitivities S_λ which can be read off on the monochromatic characteristic curves. In addition to the absolute linear construction $S_\lambda = \phi(\lambda)$, the logarithmic plot log $S_\lambda = \phi(\lambda)$ is often chosen in analogy to the extinction curves.[47] This gives, uninfluenced by absolute values, the relative spectral sensitivity distribution which is important for purposes of comparison, so that even weak sensitivity bands can be detected, these latter having an important bearing, e.g. on the discovery of relations between the constitution and the properties of compounds.

A point to be observed when discussing sensitivity curves is that since the shape of monochromatic characteristic curves is dependent on wavelength, S_λ/λ curves which have been constructed using different density indices may vary in shape. A complete description of the spectral properties of a photographic material therefore necessitates the construction of several absolute or logarithmic sensitivity curves derived at different densities (see Fig. 4).

In this connection we would refer to an instrument which was developed by *Pinoir* and *Baby*,[48] and which automatically re-records spectrograms which have been produced behind grey wedges, in the form of spectral sensitivity curves of equal densities.

Mention should also be made of the method used by *Eggert* and *Pestalozzi*[49] for measuring spectral sensitivity distribution, a method which takes into account the spectral energy distribution of the light-source. This correction

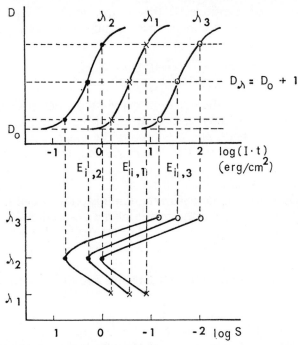

Fig. 4. Method of constructing the logarithmic spectral sensitivity curves $\log S = \phi(\lambda)$ from the monochromatic characteristic curves $D = f(\log E_\lambda)$. (Since $S = 1/E_i$; $\log S_\lambda = \log 1 - \log E_i = -\log E_i$).

is important, since if the source of energy is not standardized, readings can only be taken of the totally irradiated monochromatic radiation energy for the production of specific densities (e.g. up to $[D = D_0 + 1]$ — isodensity). The correction is made by transferring the points of intersection of the isodensities taken from the density distribution curves to a special diagram which contains (logarithmic) reciprocal curves of the energy distribution of the light-sources, constructed for different degrees of exposure (see Fig. 5). The relative spectral sensitivity is thus reproduced exactly as the reciprocal relative energy which must be absorbed at different wavelengths in order to obtain a specific density.[34]

The said method is also used by *Frieser* and *Schlesinger*[50] for measuring the

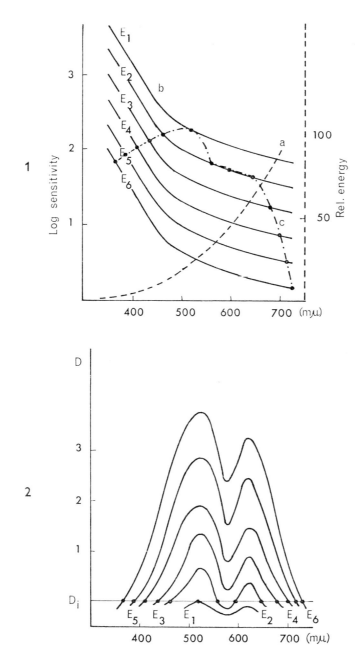

Fig. 5. Diagram for determining the spectral energy distribution. 1. Energy-gradients diagram: (a) Relative spectral energy distribution of the light source; (b) Family of curves of the reciprocal energies E_1, E_2 respectively of the log (sensitivity); (c) Relative spectral energy distribution. 2. The spectral density distribution of the various energy gradients E_1, E_2, ...

spectral sensitivity distribution of spectrally sensitized electrophotographic materials; cf. articles on xerography.[50-55]

The spectral sensitivity curves which are obtained by this method are very accurate but they are somewhat time-consuming to construct. In contrast to this, when spectrographs which are furnished with logarithmic grey wedges are used, the relative spectral sensitivity of photographic emulsion layers can be obtained as far as the infra-red region simply by making one exposure.[13,35,56] Copies of these wedge spectrograms (cf. Fig. 6) represent the relative spectral

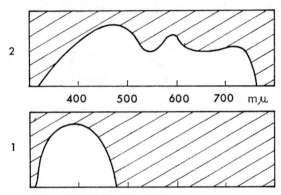

Fig. 6. Examples of wedge spectrograms (diagram). 1. Unsensitized layer. 2. Sensitized layer.

sensitivity distribution of photographic materials corresponding to a specific density.

Some information on spectral sensitivity can also be obtained from photographs of special colour charts containing colour areas as well as graduated grey areas.

1.2.2.2 *The S_λ/λ Relation in Sensitized and Non-sensitized Emulsion Layers.* The spectral sensitivity distribution in non-sensitized and sensitized photographic materials shows the following characteristics:

1. The spectral sensitivity of unsensitized photographic emulsion layers is limited to the blue and violet regions of the visual spectrum: in the case of silver iodobromide emulsions it commences at about 500 mμ and exhibits a maximum between 425 and 450 mμ. Since the spectral sensitivity follows the course of the spectral absorption of the silver halide, the large decrease in the spectral sensitivity in the long wavelength region is understandable. At 700 mμ the value of S_λ, for example, has already fallen by a factor of 10^6 to a fraction of its maximum value. At 460 mμ the decrease in the spectral sensitivity of silver bromide emulsions is interrupted by a kink which is due to a weak absorption "tail" of the silver halide in the long wavelengths, cf. *Eggert, Berg, et al.*[57-59]

2. In the case of dye-sensitized emulsions the spectral sensitivity curve

$S_\lambda = f(\lambda)$ in the region of the natural sensitivity of the silver halide ($\lambda < 500\,\text{m}\mu$) combines with that of the spectral sensitization ($\lambda > 500\,\text{m}\mu$). For these regions the following considerations apply:

a. The natural sensitivity of silver halide is often diminished by the desensitizing action of the adsorbed dye, particularly at higher dye-concentrations, and thus brings about a parallel displacement of the logarithmic sensitivity curves towards lower log S_λ values between 300 and 500 mμ.[1] The absorption of light by the dyes is not responsible for this effect.[54]

b. Sensitizing dyes endow photographic emulsions with a spectral sensitivity distribution $S_\lambda = f(\lambda)$ which corresponds approximately with one of the dye absorption curves. Depending on the sensitizer employed, photographic materials can thus be sensitized to different wavelengths in the visible region and to part of the adjacent infra-red region. Cf. in particular, Clark, et al.[55,60]

The nomenclature which is in use for this type of sensitization in the case of black-and-white materials is shown in Table 3.

TABLE 3

Colour sensitivity	Nomenclature	Wavelength region
blue sensitive	unsensitized	$\lambda \leqslant 500$ mμ
green sensitive in addition	orthochromatic	λ up to 560 mμ
green and red sensitive in addition	panchromatic (orthopanchromatic)	λ up to 700 mμ
particularly red sensitive	hyperpanchromatic	λ up to 680 mμ
infra-red sensitive	infra-chromatic	$\lambda > 700$ mμ

The relatively high blue-sensitivity of silver halides renders it more difficult to reproduce the brightness values of coloured objects to correspond with the visual impression received. Since in practice the aim is to obtain a high integral sensitivity S_i, the use of yellow filters to reduce the blue sensitivity is often dispensed with, and for this reason blue colours frequently appear too bright in comparison with the greens. With hyperpanchromatic sensitization red can also be over-accentuated.

3. With photographic colour-recording materials—cf. the reviews on this subject[61-63]—it is important for the spectral sensitivity curves of the three differently sensitized emulsion layers to be matched to the absorption curves of the image dyes. Whereas some overlapping of the sensitization is permissible in colour negatives, the sensitization maxima of colour positive materials must lie as far apart as possible. In analogy with black-and-white materials, it is imperative for colour photographic materials to have their highest sensitivity in the green region of the spectrum in order to match the sensitivity of the eye.[64]

4. Whereas ordinary sensitizing dyes enable photographs to be taken up to and far into the infra-red region, problems arise in making similar use of the

spectral region below 2200 Å owing to the strong absorption of light by the gelatin. Only two methods have proved successful for recording this short wavelength radiation (<2200 Å).

Firstly, gelatin-free Schumann plates can be employed. The disadvantage of this method lies not only in the difficulties encountered in the preparation of the plates, but also in their manipulation.

Secondly, ordinary emulsions can be "sensitized" by compounds which transform radiation of short wavelengths into fluorescent light of long wavelength which can no longer be absorbed by the gelatin. Sodium salicylate proves to be a very suitable substance for this purpose.[65] This compound excels not only in its high and constant quantum yield of fluorescence over a relatively broad region in the spectrum (600 Å to 3400 Å),[66,67] but also, amongst other things, in its stability, and because it can always be introduced in an identical manner into emulsions.[68]

5. The spectral sensitivity distribution of electrophotographic materials after sensitization is largely governed by the same laws as those which are valid for photographic emulsions: cf. *Frieser* and *Schlesinger*.[49] As in the case of silver halide, in addition to an enhancement of the sensitizing action with increase in dye concentration, desensitization also takes place, i.e. a drop in the electrophotographic sensitivity in the main lattice of the photoconductor (e.g. ZnO) has been measured. A supersensitization effect is also observed.[69]

1.2.3 MONOCHROMATIC GAMMA VALUES IN RELATION TO WAVELENGTH. As described by *Gorokhovskii*, *Farnell*, et al.,[1,70-74] the monochromatic gamma values of unsensitized and sensitized emulsions show a characteristic dependence on wavelength. The main regions to be distinguished in the γ_λ/λ curves are as follows:

1. The ultra-violet region (300 mμ–400 mμ). In this region the value of γ_λ decreases on passing to shorter wavelengths because the effective absorption of light in the lower silver halide layers of the emulsions falls with increase in the emulsion thickness Δ (see section 1.2.7).[70,72]

2. The blue-violet long wavelength region of the natural sensitivity of silver halide (400 mμ–500 mμ): in this region γ_λ is more or less constant.[1]

3. The sensitized region ($\lambda > 500$ mμ): in this region some of the values of γ_λ agree with and some of them differ from those obtained for the long wavelength region of the natural sensitivity of silver halide. Agreement of the monochromatic γ_λ values in the unsensitized and sensitized regions indicates that sensitization has no perceptible influence on the sensitivity distribution of the silver halide grains: the crystals are dyed uniformly and sensitized. The increase in γ_λ with sensitization reveals that the sensitivity of the emulsion grains is influenced non-uniformly, differences in the size of the emulsion grains being partly responsible for this effect, since dyes are adsorbed to a greater extent by small than by coarse grains (see also section 1.2.7).

It should also be noted that in describing the spectral properties of photographic materials, besides reproducing the spectral sensitivity curves $S_\lambda = \phi(\lambda)$, it is also important to draw the gamma curves $\gamma_\lambda = f(\gamma)$: for further information see *Gorokhovskii* and *Levenberg*.[1]

1.2.4 THE SCHWARZSCHILD EFFECT IN RELATION TO WAVELENGTH. The influence of wavelength on the reciprocity law is of interest in various problems in which a departure from ordinary exposure conditions arises (the recording of low light-intensities, etc.). This influence is represented graphically in the form of so-called isopaque curves which give the relation between the irradiation energy necessary to obtain a specific density [log $(I \cdot t)$] and the logarithm of the intensity or exposure time (see Fig. 7).

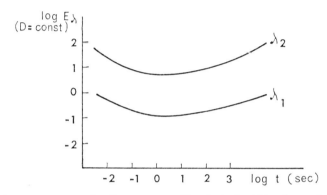

Fig. 7. Spectral isopaque curves: log $E_{\lambda[D=\text{const}]} = f(\log t)$. It should be observed that the point of inflexion of these curves ($d \log E_\lambda / d \log t = 0$) gives the optimum exposure time.

Log $E_{[D=\text{const}]}/\log t$ curves are influenced appreciably by various factors such as temperature, conditions of development, etc. The spectral composition of ultra-violet or visible radiation has, however, *no* effect on the shape of isopaque curves. These curves only become displaced parallel to one another in accordance with the spectral sensitivity of the emulsion.

Since desensitization can enhance the Schwarzschild effect[75] at low intensities, use can be made of the marked desensitizing action produced by higher concentrations of sensitizer, to influence the isopaque curves. This action, however, occurs to an equal degree at all wavelengths, so that the monochromatic log $E_D/\log t$ curves are likewise of the same shape and are displaced parallel to one another.

For the effect of infra-red sensitizers on reciprocity behaviour cf. 76.

1.2.5 NUMERICAL SPECIFICATION OF THE EFFECT OF SENSITIZATION. The different effects connected with spectral sensitization are such as to prevent sensitizing activity from being encompassed by a simple system of numbers.

For instance, when discussing the integral sensitivity of sensitized emulsion layers, two effects have to be taken into consideration. Firstly, the increase in the spectral sensitivity in the region of sensitization, and secondly, any possible decrease in the natural sensitivity of the silver halide owing to a desensitizing effect. On the other hand, in a large number of problems, particularly those arising in colour photography, the main point of interest is the width of the sensitization bands.

It is therefore necessary to specify numerically several of the effects which are linked up with sensitization. The following quantities which can be derived

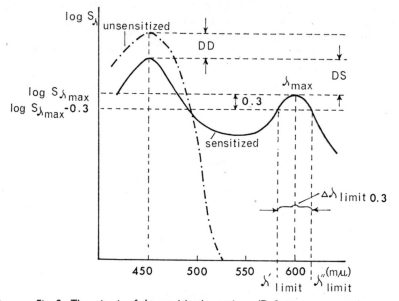

Fig. 8. The criteria of the sensitization action. (Definitions, see text.)

from the spectral sensitivity curves of *Gorokhovskii*[5] are eminently suitable for this purpose (cf. Fig. 8):

1. Degree of sensitization. This quantity corresponds to the difference between the logarithm of the spectral sensitivity at the maximum in the sensitization curve (λ_{max}) and at that of the natural sensitivity of silver halide ($\lambda = 450$ mμ) of the sensitized emulsion layers.

$$DS = \log S_{\lambda max} - \log S_{450} \qquad (6)$$

DS is usually negative and increases with the sensitizing power of a dye.

2. Degree of desensitization. This quantity is defined as the difference between the logarithm of the spectral sensitivity in the region of the natural absorption of silver halide before and after sensitization.

$$DD = \log S_{450}^{unsens.} - \log S_{450}^{sens.} \qquad (7)$$

The value of DD which is for the most part positive, increases with increase in the desensitizing action (e.g. on raising the concentration of the sensitizer).

3. Maximum sensitization λ_{max}.

4. The width of the sensitization bands $\Delta\lambda$: $\Delta\lambda$ is specified by the difference between two limiting wavelengths λ''_{limit} and λ'_{limit}, at which the spectral sensitivity $S_{\lambda limit}$ reaches a certain fraction of the maximum sensitivity $S_{\lambda max}$.

$$\Delta\lambda_{limit} = \lambda''_{limit} - \lambda'_{limit} \quad (8)$$

If the limiting wavelengths are defined for instance by

$$S_{\lambda limit} = 0.5 S_{\lambda max} \quad (9)$$

then λ'' and λ' are given by the points of intersection of the horizontal lines

$$\log S_{\lambda limit} = \log S_{\lambda max} - 0.30 \quad (10)$$

with the spectral sensitization curve.

1.2.6 THE RELATIVE QUANTUM YIELD. The relative quantum yield ϕ_r represents an important quantity for specifying the efficiency of the transfer of energy from a sensitizing dye to a silver halide. In spectral sensitometric measurements Φ_r is obtained from the ratio of the number of quanta which are absorbed by the silver halide (at 450 mμ) and by the dye, respectively (in the absorption band of the latter) and which at the time produce a specific density D:

$$\Phi_r = \frac{\text{(the number of quanta absorbed by the Ag halide)}_D}{\text{(the number of quanta absorbed by the dye)}_D} = \frac{q_{450}}{q_\lambda} \quad (11)$$

Since the energy of a quantum of light of wavelength λ [mμ] can be expressed by equation 12, then

$$\varepsilon = h \cdot \frac{c}{\lambda} = \frac{1986 \cdot 10^{-12}}{\lambda} \text{ [erg]} \quad (12)$$

a quantum number

$$q = \frac{\lambda \cdot E_\lambda}{1986 \cdot 10^{-12}} \quad (13)$$

corresponds in general to a radiation of E_λ [erg].

Making allowance for the absorption coefficient A_λ ($= J_A/J_0$) the numbers of quanta which are absorbed by the silver halide (q_{450}) and by the dye (q_λ), respectively, and which give rise to the densities which have to conform with each other, it follows therefore that

$$q_{450} = \frac{450 \cdot A_{450} \cdot E_{450}}{1986 \cdot 10^{-12}} \quad \text{and} \quad q_\lambda = \frac{\lambda \cdot A_\lambda \cdot E_\lambda}{1986 \cdot 10^{-12}} \quad (14)$$

With these values for the relative quantum yield,[38] it follows from equation 11 that:

$$\Phi_r = \frac{450 \cdot E_{450} \cdot A_{450}}{\lambda \cdot E_\lambda \cdot A_\lambda} \qquad (15)$$

The relative quantum yield is closely related to the degree of sensitization DS. If Φ_r is specified by the spectral sensitivity in the regions of maximum sensitization ($S_{\lambda\max}$) and of silver halide absorption (S_{450}),[5] then equation 15 can be replaced by:

$$\Phi_r = \frac{S_{\lambda\max}/A_{\lambda\max}}{S_{450}/A_{450}} \quad \text{or} \quad \Phi_r = \frac{S_{\lambda\max}}{S_{450}} : \frac{A_{\lambda\max}}{A_{450}} \qquad (16)$$

In equation 16 the ratio of the absorption coefficients $A_\lambda = A_{\lambda\max}/A_{450}$ corresponds to the amount of dyeing produced by the dye in the long wavelength region of the spectrum. Expressed logarithmically, this equation reads

$$\log \Phi_r = (\log S_{\lambda\max} - \log S_{450}) - \log A_\lambda \qquad (17)$$

and by introducing the degree of sensitization DS in accordance with equation 6, we have:

$$\log \Phi_r = DS - \log A_\lambda \qquad (18)$$

According to this, the relative quantum yield Φ_r can also be determined from the spectral sensitivity curve, taking A_λ into consideration. However, in neither of the equations 15 and 18 has the influence of desensitization been taken into account.

Measurements of the relative quantum yields give values of up to an order of 1 in magnitude.[43] However, Φ_r by no means remains constant throughout the entire region of sensitization. It varies particularly when the dyes are adsorbed to the silver halide in various states of aggregation:[43,77] to give an example, in general Φ_r is smaller in the case of an H-sensitization band than when the dye is present in the monomeric M state. As soon as J-aggregates are present (sensitization of the 2nd order) Φ_r can reach very high values.[5] The additional substances present in an emulsion layer (surface active organic compounds, colour couplers, etc.) also exert an influence on Φ_r.[78,79] The removal of excess bromide ions also increases Φ_r and therefore the sensitizing action.[79,79a]

These few examples show that measurements of the relative quantum yield can furnish important practical information on the mechanism of spectral sensitization.

1.2.7 THE INFLUENCE OF DYE-SENSITIZATION ON THE CHARACTERISTIC CURVE. Dye-sensitization not only produces a displacement in the position of the characteristic curve on the log E axis—a position which is important for the sensitivity—but it also influences the form of the $D/\log E$ relation.

Since—apart from the effect of vacant sites—the absorption of light by the emulsion grains plays the most important part in sensitivity, the sensitivity of unsensitized and sensitized emulsions in relation to the spectrum is attributable in many cases to changes in the absorption of light by the grains and to fluctuating attenuations in the intensity of the light within the emulsions, such fluctuations being dependent on wavelength. In particular, when the relative quantum yield is satisfactory spectral sensitization has the same effect as an increase in the absorption at the grain surface. The relation between the sensitivity S and dye-sensitization can therefore, according to *Farnell*,[79b] be simplified by means of equation 19 (in which ε is the extinction coefficient of the sensitizer, which is dependent on wavelength, and d is the thickness of the dye):

$$\log S = \log (1 - e^{-\varepsilon d}) + \text{const.} \qquad (19)$$

On the other hand, various effects are responsible for the influence of spectral sensitization on the $D/\log E$ relation, i.e. for the gamma value of sensitized emulsion layers in relation to wavelength.

According to *Farnell*[70] changes in the unexposed emulsion density Δ which is such an important factor in determining the behaviour of unexposed multi-layer emulsions can lead to irregularities in the gamma values. The unexposed emulsion density Δ is given by the exponential drop in intensity inside the emulsions and by the logarithm of the ratio of the intensity of light which is effective in the top and bottom layers of grains, so that Δ bears a reciprocal relation to the gamma value. When the absorption of light by the uppermost layer of grains is low, a small Δ value of a multilayer emulsion postulates a high gamma value: an increase in Δ such as occurs for instance in silver halide on passing to short waves (up to 300 mμ) is, on the other hand, linked with a decrease in γ. The minimum values of gamma which were measured by *Farnell*[70] in the region of the absorption-maxima of the sensitizers can, accordingly, be explained by the occurrence of high Δ values in the places of maximum dye absorption, since the upper grain layers produce a particularly great decrease in the intensity of the light on account of the high extinction of the dye.

In contrast to this, any increase in gamma which is observed in passing from the region of the natural absorption of the silver halide to the area of sensitization cannot be attributed to effects which are linked up with changes in Δ. In spite of an increase in γ, Δ remains constant during this transition. The increase in γ in this case might probably be due to an adjustment of the sensitivities of different grain-size classes brought about by the sensitization.[5] Since the grains of different sizes are usually widely distributed throughout the emulsions, and therefore the sensitivity distribution is non-uniform, this effect can often be detected. The small grains in these emulsions do indeed adsorb relatively more dye than the coarse crystals, so that it is primarily the threshold energy of the small grains which benefits. Thus, as the result of sensitization, a relatively sensitive but non-uniform emulsion layer which has previously been assigned a low gamma value—cf. Fig. 9—acquires the nature of a sensitive

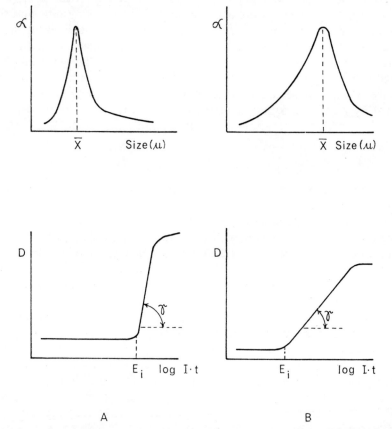

Fig. 9. Diagram of the connection between grain size distribution, sensitivity $S = k/E_i$ and gamma value γ. A: Fine emulsion crystals. B: Coarse emulsion crystals.

emulsion of uniform sensitivity distribution. These emulsions have in common a high gamma value, whereby it is immaterial whether the uniform distribution of the grains is present from the commencement or whether the uniform sensitivity distribution is brought about by sensitization.

References

[1] GOROKHOVSKII, YU. N. and LEVENBERG, T. M., *General Sensitometry*, Focal Press (London, New York) 1965.

[2] TUPPER, J. L., *The Theory of the Photographic Process*, 3rd edn., edited by C. E. K. MEES and T. H. JAMES, p. 409, Macmillan (New York, London) 1966.

[3] LOBEL, L. and DUBOIS, M., *Sensitometry*, Focal Press (London, New York) 1955.
[4] KUHN, G., *Ullmanns Encyklopädie der technischen Chemie* (*Ullmanns Encyclopedia of Industrial Chemistry*), 3rd edn., p. 686, Urban u. Schwarzenberg (Munich-Berlin) 1962.
[5] GOROKHOVSKII, YU. N. *Spectral Studies of the Photographic Process*, Focal Press (London) 1965.
[6] EGGERT, J. and VON WARTBURG, R., *Z Electrochem*, **61**, 693 (1957).
[7] POHL, R. W., *Einführung in die Physik*, 3. Band Optik, (*Introduction to Physics*, Vol. 3, Optics), p. 320 et seq., Springer (Berlin-Göttingen-Heidelberg) 1948.
[8] SELIGER, H. H. and MCELROY, W. D., *Light: Physical and Biological Action*, p. 5 et seq., Academic Press (New York and London) 1965.
[9] BAUER, G., *Measurement of Optical Radiations*, Focal Press (London) 1962.
[9a] TOSIIIO, ABE, *Nippon Shashin Gakkai Kaishi*, **28**, 83 (1965).
[10] EGGERT, J. and VON WARTBURG, R., *Z Elektrochem*, **59**, 353 (1955).
[11] TAYLOR, A. H. and KERR, G. P., *J. opt. Soc. Amer.*, **31**, 3 (1941).
[12] WILLIAMS, F. C. and GRUM, F., *Photogr. Sci. Eng.*, **4**, 113 (1960).
[13] MEES, C. E. K. and WRATTEN, S. H., *Brit. J. Photogr.*, **54**, 384 (1907).
[14] JONES, L. A. and SANDVIK, O., *J. opt. Soc. Amer.*, **12**, 401 (1926).
[15] EVANS, C. H., *J. opt. Soc. Amer.*, **30**, 118 (1940).
[16] MORRISON, C. A. and HOADLEY, H. O., *P.S.A. Journal*, **16B**, 64 (1950).
[16a] JELLINEK, G. and SCHEDEL, E., *Z. Instr.*, **73**, 179 (1965).
[17] HARTMAN, A. W., *Rev. Sci. Instrum.*, **36**, 31 (1965).
[18] CALLIER, A., *Z. wiss Photogr.*, **7**, 257 (1909).
[19] EGGERT, J. and KÜSTER, A., *Kinotechnik*, **16**, 127 (1934); *Veröffentlichungen des wissenschaftlichen Zentral-Laboratoriums der photographischen Abteilung AGFA* (*Publications of the Photographic Department of the AGFA Central Scientific Laboratory*), Vol. IV, p. 50, Hirzel (Leipzig) 1935.
[20] MORRISON, C. A., *J. opt. Soc. Amer.*, **42**, 90 (1952).
[21] OWEN, R. E. and DAVIES, E. R., *Photogr. J.*, **74** (N.S. 58), 463 (1934).
[22] GOLDBERG, E., *Der Aufbau des photographischen Bildes* (*The Structure of the Photographic Image*), p. 100, Verlag Knapp (Halle/Saale) 1925.
[23] THIELS, A., *J. photogr. Sci.*, **8**, 236 (1960).
[24] LUTHER, R., *Brit. J. Photogr.*, **57**, 664 (1910).
[25] HIGSON, G. I., *Photogr. J.*, **61** (N.S. **45**), 93 (1921).
[26] MÜLLER, R., *Mitt ForschLab Agfa-Gevaert Leverkusen-München*, **4**, 346 (1964).
[27] MIKHAILOV, V. YA., *Fortschr. wiss Photogr.* (*UdSSR*), **10**, 102 (1964).
[28] IVANOVA-OGERELIEVA, E., *IVth Conf. on sci. appl. Phot.* Budapest, 1963, 155, quoted from *Chem. Zbl.* 33/3222 (1966).
[29] V. ANGERER, E., *Wissenschaftliche Photographie* (*Scientific Photography*), Akad. Verlagsgesellschaft Geest u. Portig (Leipzig) 1953.
[30] LUTHER, R., *Trans. Faraday Soc.*, **19**, 340 (1923).
[31] JONES, L. A., *J. Franklin Inst.*, **227**, 297 and 497 (1939).
[32] JONES, L. A. and NELSON, C. N., *J. opt. Soc. Amer.*, **30**, 93 (1940).
[33] NELSON, C. N. and SIMONDS, J. L., *J. opt. Soc. Amer.*, **46**, 324 (1956).
[34] VIETH, G. and HEILAND, W., *Photogr. Korr.*, **97**, 99 (1961).
[35] BAKER, T., *Photographic Emulsion Technique*, 6th edn., pp. 280, 290, Chapman and Hall (London) 1949.
[36] WEST, W. and CARROLL, B. H., *J. chem. Phys.*, **15**, 529 (1947).
[37] BRÜNNER, R., GRAF, A. and SCHEIBE, G., *Z. wiss Photogr.*, **53**, 214 (1959).
[38] WEST, W. and CARROLL, B. H., *The Theory of the Photographic Process*, 3rd edn., edited by C. E. K. MEES and T. H. JAMES, p. 233, Macmillan (New York, London), 1966.
[39] STERN, O. and VOLMER, M., *Phys. Z.*, **20**, 183 (1919).
[40] WEST, W. and CARROLL, B. H., *J. chem. Phys.*, **19**, 417 (1951).
[41] DÄHNE, S., *Z. wiss Photogr.*, **59**, 113 (1965).
[42] EGGERT, J. and BILTZ, M., *Veröffentlichungen des wissenschaftlichen Zentral-Laboratoriums der photographischen Abteilung AGFA* (*Publications of the Photographic Department of the AGFA Central Scientific Laboratory*), Vol. VI, p. 23, Hirzel (Leipzig), 1939; *Trans. Faraday Soc.*, **34**, 892 (1938).
[43] BREIDO, I. I. and GOROKHOVSKII, YU. N., *Zh. Fiz. Khim.*, **17**, 57 (1943).
[44] SPENCE, J. and CARROLL, B. H., *J. phys. Chem.*, **52**, 1090 (1948).
[45] BORGINON, H. and VAN VEELEN, G. F., *J. photogr. Sci.*, **13**, 273 (1965).

[46] LEERMAKERS, J. A., CARROLL, B. H. and STAUD, C. J., *J. chem. Phys.*, **5**, 893 (1937).
[47] PESTEMER, M., *Anleitung zum Messen von Absorptions-spektren im Ultraviolett und Sichtbaren (Introduction to the Measurement of Absorption Spectra in the Ultraviolet and the Visible)*, Georg Thieme Verlag (Stuttgart) 1964.
[48] PINOIR, R. and BABY, P., *Sci. Ind. Photogr.*, **22** (2), 88 (1951).
[49] EGGERT, J. and PESTALOZZI, H., *Z. Elektrochem.*, **65**, 50 (1961).
[50] FRIESER, H. and SCHLESINGER, M., *Photogr. Korr.*, **101**, 70 and 133 (1965).
[51] DESSAUER, J. H., MOTT, G. R. and BOGDONOFF, H., *Schweiz. Photordsch.*, 1–3, 1957 (Translation A. BÜCHLER and J. EGGERT); *Photogr. Eng.*, **6**, 251 (1955).
[52] HAUFFE, K., *Angew. Chem.*, **72**, 730 (1960); *J. photogr. Sci.*, **10**, 321 (1962).
[53] MEIER, H., *Die Photochemie der organischen Farbstoffe (The Photochemistry of Organic Dyes)*, Springer (Berlin) 1963.
[54] LANGSTON, D. J., *Tappi*, **49**, 16 (1966).
[55] HOEGL, H., *J. phys. Chem.*, **69**, 755 (1965).
[56] CLARK, W., *Photography by Infrared*, John Wiley & Sons (New York) 1947.
[57] EGGERT, J., BILTZ, M. and KLEINSCHROD, F. G., *Z. wiss Photogr.*, **39**, 140 and 155 (1940).
[58] EGGERT, J. and KLEINSCHROD, F. G., *Z. wiss Photogr.*, **39**, 165 (1940).
[59] BERG, W. F., *Phys. Soc. Rep. Progr. Physics*, **11**, 264 (1946/47).
[60] NÜRNBERG, A., *Infrarot Photographie (Infrared Photography)* VEB Wilhelm Knapp Verlag (Halle/Saale) 1957.
[61] EGGERT, J. and HEYMER, G., *Naturwissenschaften*, **25**, 689 (1937).
[62] EGGERT, J., *Z. Elektrochem.*, **56**, 712 (1952).
[63] WEISSBERGER, A., *The Theory of the Photographic Process*, 3rd edn., edited by C. E. K. MEES and T. H. JAMES, p. 382, Macmillan (New York) 1966.
[64] BARCHET, H. M., *Chemie photographischer Prozesse (Chemistry of Photographic Processes)*, Akademie Verlag (Berlin) 1965.
[65] ALLISON, R. and BURNS, J., *J. opt. Soc. Amer.*, **55**, 574 (1965).
[66] JOHNSON, F. S., WATANABE, K. and TOUSEY, R., *J. opt. Soc. Amer.*, **41**, 702 (1951).
[67] ALLISON, R., BURNS, J. and TUZZOLINO, A. J., *J. opt. Soc. Amer.*, **54**, 747 (1964).
[68] ALLISON, R., BURNS, J. and TUZZOLINO, A. J., *J. opt. Soc. Amer.*, **54**, 1381 (1964).
[69] INOUE, E. and YAMAGUCHI, T., *Bull. chem. Soc. Japan*, **36**, 1573 (1963).
[70] FARNELL, G. C., *J. Photogr. Sci.*, **2**, 145 (1954).
[71] FARNELL, G. C., *J. Photogr. Sci.*, **7**, 83 (1959).
[72] AMSTEIN, E. H., *J. Soc. chem. Ind., London*, **63**, 172 (1944).
[73] SILBERSTEIN, L., *J. opt. Soc. Amer.*, **32**, 326 (1942).
[74] WEBB, J. H., *J. opt. Soc. Amer.*, **38**, 27 (1948).
[75] TUTIHASI, S., *J. opt. Soc. Amer.*, **45**, 15 (1955).
[76] SHVARTS, V. M. and DOKUCHAEVA, Z. P. *Zh. nauch. prikl. Fotogr. Kinem.*, **10**, 261 (1965).
[77] LEERMAKERS, J. A., *J. chem. Phys.*, **5**, 889 (1937).
[78] GOROKHOVSKII, YU. N. and GRATSIANSKAYA, Z., *Zh. nauch. prikl. Fotogr. Kinem.*, **2**, 421 (1957).
[79] NATANSON, S. V., *Zh. nauch. prikl. Fotogr. Kinem.*, **8**, 3462 (1963).
[79a] GROSS, L. G., *Zh. nauch. prikl. Fotogr. Kinem.*, **10**, 250 (1965).
[79b] FARNELL, G. C., *The Theory of the Photographic Process*, 3rd edn., edited by C. E. K. MEES and T. H. JAMES, p. 72, Macmillan (New York, London) 1966.

II. SENSITIZING DYES

2.1 *The Relation between the Constitution of a Dye and the Region of Sensitization*

2.1.1 THE POSITION OF MAXIMUM SENSITIZATION IN THE SPECTRUM. The discovery of spectral sensitization by *Vogel*[80] as long ago as 1873 indicated the path which had to be followed in order to bring the visible and infra-red regions of the spectrum within reach of photography. The first experiments on sensitization which also included Becquerel's interesting observation[81] of the sensitizing power of chlorophyll in photosynthesis showed that the spectral distribution of sensitized photographic sensitivity agreed with the fundamental photochemical law of *Grotthus-Draper*[82] in that it was dependent on the position of the region in which the dye absorbed.

The first sensitizers encompassed only a part of the region of the visible spectrum, namely that up to 600 mμ. Subsequently, however, the discovery and synthesis by *Homolka, Brooker, Scheibe, Dieterle, Mills, Riester, Heseltine* and others[83-89] of suitable sensitizing dyes enabled the photographic spectrum to be extended not only to cover the entire visible region of the spectrum, but also far into the infra-red region—up to and beyond 1300 mμ. The spectral sensitivity of sensitized emulsions coincides approximately with the position of the absorption spectrum of the dye measured in solution,[89a] and thus the connection between the region of sensitization and the constitution of the sensitizer can be attributed to the relations existing between the constitution of the dye and the absorption of light by the latter.

2.1.2 THE DISPLACEMENTS IN ABSORPTION CAUSED BY ADSORPTION. When discussing the last-named relation it must, however, be borne in mind that absorption spectra, and therefore sensitization spectra, may differ from the spectra measured in solution, due to the adsorption of dye molecules to the silver halide grains.

Firstly, when there is only slight coverage of the silver halide grains by the dye, the absorption bands of sensitizing dyes are seen to be situated 20–30 mμ farther into the long wavelengths than those measured in solution, this displacement being partly due to the effect of dispersion forces, and partly to electrostatic interaction between the dye and the substratum.[90-92]

Secondly, with certain dyes which are adsorbed in fairly high concentrations, the position of the absorption band may undergo hypsochromic displacement (i.e. towards the short wavelengths) or bathochromic displacement (i.e. towards the long wavelengths). Here we have an effect which is also observed in aqueous dye solutions and which is due to aggregation of the dye, which latter is dependent on the concentration of the dye in solution. This aggregation, which has been thoroughly studied and explained by *Scheibe et al.*,[93-96] passes through several intermediate states which are indicated in the absorption spectrum by characteristic absorption bands. Whilst with many dyes only some of these characteristic bands can be detected, with others—e.g. 1,1′-diethyl-2,2′-cyanine (1)—all the intermediate states are observed when the concentration is raised, e.g. from 10^{-5} mols/litre to 10^{-2} mols/litre.

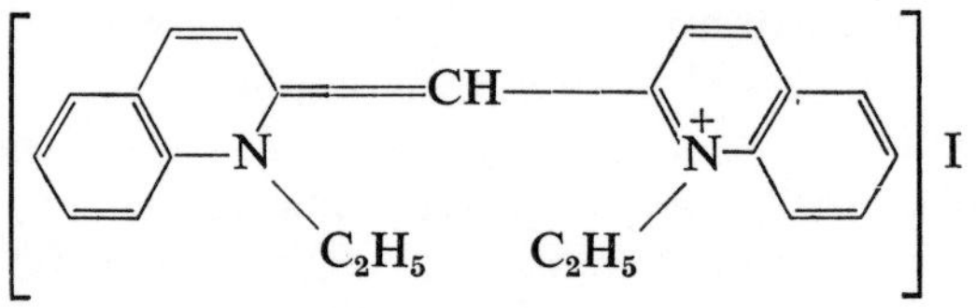

[1]

The following bands and states respectively—in addition to the monomeric bands—are distinguished:

1. *D-bands.* These bands which are caused by dimerization of the dye molecules are greatly dependent on concentration and temperature.[97,98] They are observed in many classes of dyes (cyanines, merocyanines, rhodamines, thiazines, triphenylmethane dyes, etc.) and have two characteristic components, one of which undergoes hypsochromic displacement in relation to the monomeric band.[97,99,100,101,153]

2. *H-bands.* These bands which lie farther in the short wavelength region than do the monomeric bands, can occur in aqueous solution on raising the concentration of the latter (and may be preceded by the formation of a D-band). H-bands are also observed in many cases during the adsorption of a dye to the silver halide grains, and in the opinion of *Leermakers, Carroll* and *Staud*[46] are, like the J-state, attributable to (unorientated) aggregates of dye molecules; see also[101a]. *Padday* and *Wickham*,[236a] *McKay* and *Hillson*[101b] however, assume that it is not intermolecular interaction between the aggregated molecules which is responsible for the formation of the H-bands in the spectrum of the dye, but interaction between the chromophore groups and the electrical field at the surface of the silver halide and the counter ions of the dye respectively.

3. *J-bands* are bands which are situated 50–100 mμ farther in the long wavelengths and are characterized by their high intensity and sharpness. These bands are attributed by *Scheibe et al.*[93,94,98,102-104] to the formation of a reversible state of dye-aggregation. The latter reverts to the monomeric state on raising the temperature, and, amongst other things, is characterized by its fluorescent power, streaming anisotropy and the parallel arrangement of the dye molecules,[53,103-107] which can already form three molecules.[108] This state of dye aggregation is observed chiefly in the case of readily soluble cyanines. However, other dyes (e.g. rhodamines) also give indications of J-states in certain

conditions.[97] Above all it should be noted that the J-state is most liable to occur in gelatin solutions and on adsorption of the dye to the silver halide.[38] Many dyes are capable of forming J-aggregates by means of adsorption, this process being influenced by the nature of the crystal surface,[109,110,111] the type of gelatin,[112,112a] the iodide and water content of the emulsion[113–116] or the presence of readily volatile solvents.[117] These aggregates are not formed in solutions, even at high concentrations (see also section 2.2.5). In this connection it should be mentioned that the spectral absorption of adsorbed cyanine dyes, i.e. the formation of H- and J-states, depends on the silver halide substratum. *Borginon et al.*[110,111,114] attribute this to a variation in the power of substrata to adsorb aggregates of similar structure. *Paddy's*[236] measurements show, however, that the actual state of the dye itself is determined by the substrata. Accordingly it is not the surface concentration of the adsorbed dyes, but the formation of complexes or a kind of ion exchange between the adsorbed dye and the silver halide which should be held responsible for the fact that the formation of H- and J-states is dependent on the substratum. The relation between photographic sensitivity and wavelength follows the same path as the J-bands, the corresponding sensitization effect being known as sensitization of the 2nd order. Sensitization bands of the 2nd order are often characterized by their sharp definition—particularly in the case of supersensitization. However, in contrast to the bands occurring in other regions of the spectrum, differences in the photographic properties (gamma value, etc.) are also observed. This will be further discussed in section 2.5. The most significant feature is that the efficiency of the transfer of energy from the dye to the silver halide, as derived from measurements of Φ_r, is dependent on the state of aggregation of the dye.[79]

2.1.3 THE THEORY OF THE ABSORPTION OF LIGHT BY DYES.

2.1.3.1 *The Electron Gas Model.* Since the said displacements in absorption group themselves respectively around the long wavelength monomeric dye bands, approximately a direct relation can be said to exist between the absorption spectrum of dye molecules and their sensitization spectra. A theory of dye absorption thus enables suitable dyes to be selected for sensitizing a particular region of the spectrum.[118]

The development of a theory for the light-absorption of sensitizers is facilitated by the fact that most sensitizers belong to the polymethine class of dyes. This is because the colour of the dyes which are described by the general formula

$$X - \acute{C} = (\acute{C} - \acute{C} =)_n Y$$

(X, Y are heterocyclic atoms which are usually built into the ring systems; n is a whole number including nought) is due to electron gas clouds which are capable of resonating and which are comprised of the π-electrons of the auxochrome (i.e. of the coordinatively unsaturated atoms or groups of atoms such as N, O or S) and those of the methine chain. The π-electrons may therefore be regarded as the valency electrons of a metal, these electrons being free to move about within the potential walls set up by the atomic framework of the molecules,[119,120] so that it is possible to use the electron gas model developed by *H. Kuhn*[121-123] to describe the electron states which are so important for absorption.

To this end the discrete energy levels and the characteristic functions, respectively, of a π-electron held inside the potential well of a molecule—cf. Fig. 10

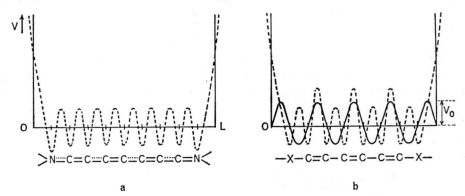

Fig. 10. Diagram of the flow of the potential energy of an electron inside a dye molecule. (a) Symmetrical end groups. (b) Unsymmetrical end groups.

—are calculated from the *Schrödinger* equation or by means of the reflection oscillator method.

With the last-named method the π-electron system is as it were regarded as a vibrating rod in which, according to the laws of vibration, stationary waves are only set up when there are vibration nodes as the ends of the system. The following equation holds for the possible wavelengths

$$\lambda_n = \frac{2L}{n}, \tag{20}$$

in which L is the length of the π-electron chain and n is a whole number. If λ_n is equated to a de Broglie wave

$$\lambda_n = \frac{h}{m \cdot u} \tag{21}$$

then for the energy $E = \frac{m}{2} u^2$ of the discrete energy levels we obtain

$$E_n = \frac{h^2}{8mL^2} \cdot n^2 \ (n = 1, 2, \ldots) \tag{22}$$

in which h denotes the Planck quantum of action ($6\cdot62 . 10^{-27}$ erg/sec), m, the mass of the electrons ($m = 9\cdot107 . 10^{-28}$ g) and c, the velocity of light ($c = 2\cdot998 . 10^{10}$ cm/sec). The *Schrödinger* equation leads to the same result. Cf. for example, 53.

It should also be noted that the density distribution of the electrons which are present in states of energy of E_n can be derived from the form of the function ψ_n of a simple vibrating system. Since the density distribution of the electrons is given by $e|\Psi_n^2|dv$, it is possible by means of the Ψ_n^2-formation of the wave function Ψ_n

$$\Psi_n = \sqrt{\frac{2}{l \cdot N}} \sin \frac{\pi \cdot s}{l \cdot N} \cdot n \quad (n = 1, 2, \ldots) \tag{23}$$

to give a qualitative description of the distribution of the electrons along the C-chain—as is shown in Fig. 11. In equation 23, l represents the distance

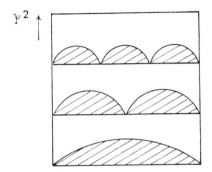

Fig. 11. The electron density distribution of a linear conjugated double bond system.

between two C atoms or between C and nitrogen, and N is the number of π-electrons. s is the distance of a considered position in the electron system counted from the commencement of the chain.

In calculating the 1st absorption bands it is, however, important for the molecular orbits to be filled, from the lowest state upwards, with two π-electrons respectively in observance of the *Pauli* principle. Thus, N π-electrons possess $n = N/2$ states, so that the highest energy level occupied is denoted by $E_{N/2}$ and the first free level by $E_{N/2+1}$. The difference in energy corresponding with the excitation energy of these two terms, and at the same time taking equation 22 into consideration, is given by

$$\Delta E = E_{(N/2)+1} - E_{N/2} = \frac{h^2}{8mL^2}(N+1) \tag{24}$$

Since the following general relation exists between the absorption bands and the excitation energy:

$$\Delta E = h\nu = h \cdot \frac{c}{\lambda} \qquad (25)$$

the position of the 1st absorption band will be given by

$$\lambda_{\max} = \frac{h \cdot c}{\Delta E} = \frac{8mc}{h} \cdot \frac{L^2}{N+1} \qquad (26)$$

The length L of the conjugated double bond system contained in equation 26 is calculated from the number of π-electrons N and the distance l of the atoms in accordance with

$$L = N \cdot l \qquad (27)$$

the number of π-electrons N in polymethine chains with nitrogen atoms at each end being given by:

$$N = \text{(the number of C atoms in the chain)} + 3 \qquad (28a)$$

and $\qquad N = \text{(the number of C and N atoms)} \quad + 1 \qquad (28b)$

Equations 28a and 28b can be derived from the rule governing standard polymethine dyes. This rule states that a polymethine chain consisting of sp^2 hybridized carbon atoms contains an *odd* number of members in the chain between the atoms or groups of atoms at its two ends. Firstly, namely by the sp^2 hybridization, i.e. by the coalescence of the three valency electrons ($2s$, $2p_x$, $2p_y$) of a C atom, a $2p_z$ electron which is situated perpendicularly to the sp^2-plane is made available at the time to the π-electron system by a carbon atom.

On the other hand, three π-electrons must formally be added to the π-electron system of polymethine dyes which terminate in two substituents with free electron-pairs (atoms of the 5th or 6th main group of the periodic system) (equation 28b). Thus, in general, the number of π-electrons N in polymethine dyes is one greater than the number of members p in the polymethine chain. However, by choosing special groupings of atoms at the ends of the chains—as in the cyanines containing phosphorus—the number of π-electrons can also be decreased in comparison with the total number of atoms p, by one electron, so that, according to *Dähne*,[123a] the number of π-electrons in polymethine dyes can, in a more restricted sense, generally be expressed by

$$N = p \pm 1 \quad [p \text{ is an odd number}] \qquad (28c)$$

The connection between the long wavelength absorption band λ_{max} and the number of π-electrons N can be derived from equations 26 and 27, and gives the relation

$$\lambda_{max} = \frac{8mc}{h} \cdot \frac{N^2 \cdot l^2}{N+1} \tag{29}$$

After inserting the numerical values (the mean distance between the bonds l being equal to $1 \cdot 39 \cdot 10^{-8}$ cm), we have

$$\lambda_{max} = 637 \cdot \frac{N^2}{N+1} \, [\text{Å}] \tag{30}$$

If N be substituted by the number of double bonds $d - N = (d+1) \cdot 2 -$ then we obtain

$$\lambda_{max} = 127 \cdot \frac{(d+1)^2}{d+\frac{3}{2}} \, [m\mu] \tag{31}$$

The electron gas theory enables the absorption bands of longest wavelength to be calculated without the aid of any empirical data. The principle advantage of the theory is that not only does it calculate accurately the experimentally observed bathochromic displacement of the absorption maximum when the system of double bonds is enlarged, but it also embraces other changes in absorption due to structure.

This holds also for the calculation of the absorption of light by polymethine radicals[123b] which contain an *even* number of methine atoms between the two ends of the chains and therefore, according to equation 28, have an *odd* number of π-electrons:

$$N = p \pm 1 \, [p \text{ is an even number}] \tag{28d}$$

Thus it is not a solitary electron which is responsible for the absorption of light by these radical-ions[123c] but the overall electron distribution which can be encompassed by the electron gas model. The slight difference between the measured values of the half-stage potentials at the anode and cathode in the case of radicals (Wurster's blue)[123d] agrees with these findings. The highest electron level in the radical is indeed only half occupied, so that its energy relations correspond both with the highest occupied electron level present in the dyes (Δ half-stage potential at the anode) and with the lowest free electron level (Δ half-stage potential at the cathode). For definitions see Chap.VI, 2, 3A). The half-occupied level thus fixes simultaneously the energy of the anodic and cathodic half-stage potentials. The absorption of light cannot have any influence on these potentials, which, in radicals, are for the most part equal in value, because the transition from the highest double-occupied to the half-occupied level proceeds at the longest wavelength which can be encompassed by the electron gas model.[41,123e]

In spite of the consistent use of the electron gas model for calculating the absorption of light by polymethine dyes, it will, of course, be understood that polymethine radicals differ from one another in the reactivity which they promote in these dyes. For instance, the polymethine radicals in which N is an odd number are powerful desensitizers.[123f,123g]

2.1.3.2 *The Relation between Light-absorption and Constitution.* The paramount importance of the relation between light-absorption and constitution for the synthesis of suitable sensitizers for specific spectral regions is readily understandable. The use of the following rules, which can partly be explained by the electron gas model, enables a number of predictions of practical importance to be made regarding the position of the maximum absorption.

1. *The 100 mμ rule*[123–126] postulates that the introduction of new vinyl groups (—CH=CH—) which respectively correspond to two new π-electrons, brings about a steady displacement of about 100 mμ per double bond towards longer wavelengths of the absorption spectrum of symmetrical polymethine dyes. As shown by the comparison of the experimental and theoretical values obtained for the first absorption maxima given in Table 4 for the symmetrical carbocyanine dye [2]

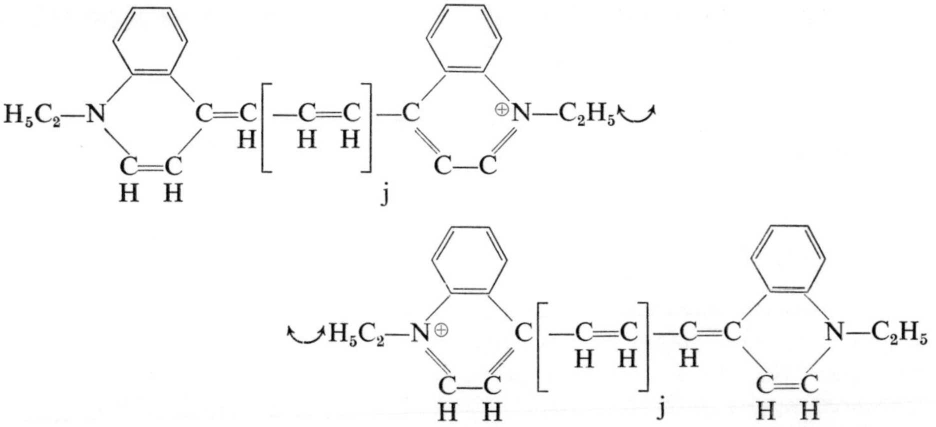

the 100 mμ rule is covered by the electron gas theory.

It can also be seen from Table 4 that the maximum absorption can be displaced as far as the infra-red region of the spectrum by building a large number of methine groups into the molecule. For instance, the maximum sensitization of benzothia-undecacarbocyanine, which contains 11 methine groups between the nuclei, is at 1·05 μ.

When aromatic nuclei, such as those occurring for instance in the cyanines

TABLE 4

Number of π-electrons (N)	Number of double bonds d	Number of vinyl-groups j	λ_{max} theoretical Å	λ_{max} measured[124] Å
10	4	0	5765	5900
12	5	1	7060	7100
14	6	2	8340	8200
16	7	3	9590	9300

(Pinacyanol, etc.) are present, allowance must be made for the influence of the benzene nuclei when making the calculations, otherwise the absorption bands λ_{max} would appear to occur at too short a wavelength. In these compounds the changes in potential at the ends of the chains can no longer be simplified in terms of a vertical increase, since the polarizability of the π-electrons in the aromatic nuclei brings about a relatively slow increase in the potential energy at the end of the conjugation system which is responsible for the absorption of light. Kuhn[121,123] makes allowance for this influence by extending the length L of the conjugation system by 2/3 of the distance between two atoms l

$$L = N \cdot l + \tfrac{2}{3}l = l \cdot (N + \tfrac{2}{3}) \qquad (32)$$

Equation 29 therefore has to be replaced by

$$\lambda_{max} = \frac{8mc}{h} \cdot \frac{l^2 \cdot (N + 2/3)^2}{N + l} \qquad (33)$$

With many compounds further lengthening of the chain may be necessary in order to reproduce accurately the experimental findings.

2. *On substitution in the conjugated chain*—e.g. when one of the —CH= methine groups is replaced by an N atom, or a nitrile group is introduced—hypsochromic or bathochromic displacements of up to an order of magnitude of 100 mμ occur in the absorption.[127–130]

For instance, the replacement of the ring-closing carbon atom in Acridine orange NO[3] (λ_{max} = 491·5 mμ) by a nitrogen atom results in Amethyst violet [4] (λ_{max} = 590), a dye which absorbs at almost 100 mμ farther into the long wavelength region.

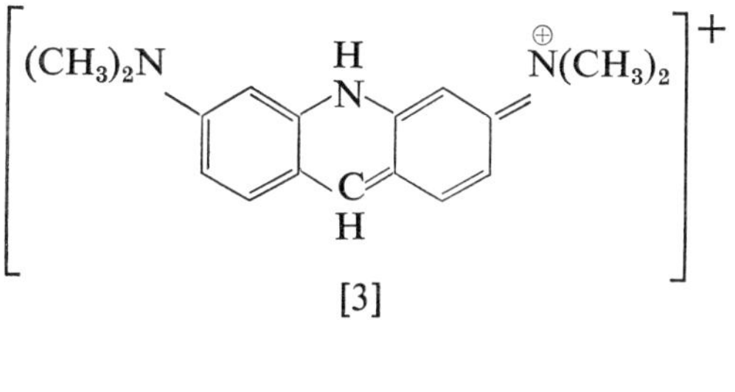

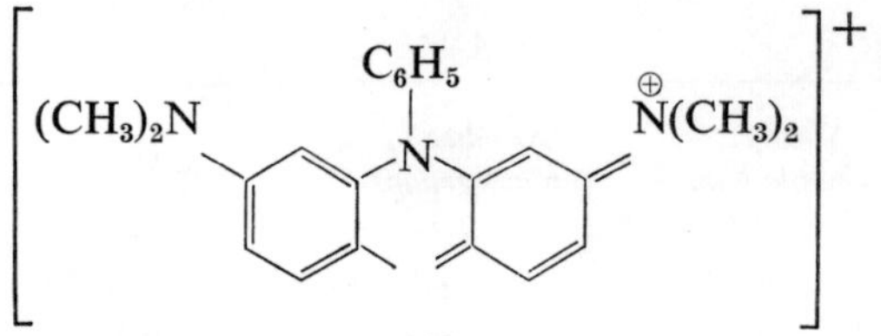

$$\left[(CH_3)_2N - \underset{\underset{}{}}{\overset{C_6H_5}{\underset{|}{N}}} - \overset{\oplus}{N}(CH_3)_2 \right]^+$$

[4]

The absorption of Astraphloxine [5] at 544 mμ is likewise displaced towards the long wavelengths when a nitrile group is substituted in the meso-position: the substitution product [6] absorbs at 600 mμ.

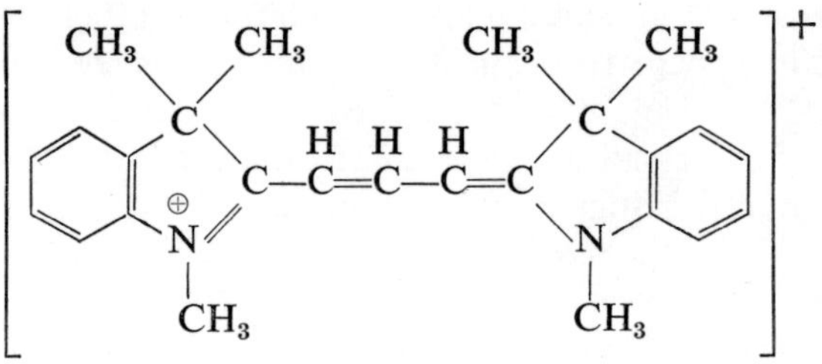

[5]

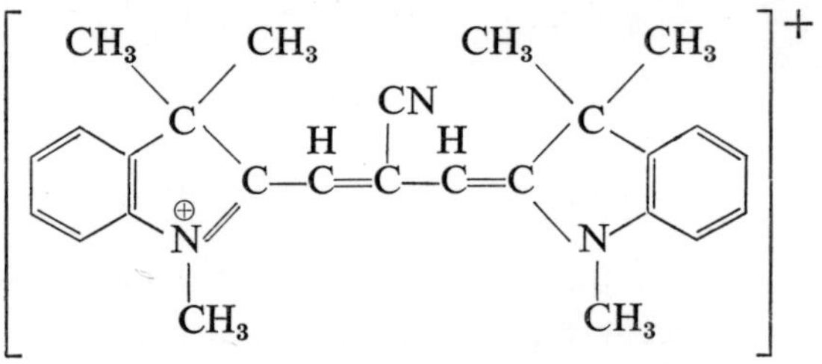

[6]

Disturbance of the mobility of the π-electrons in the conjugated chain may be regarded as the actual cause of this effect because an N-atom or a nitrile group possesses a greater electron affinity than that of a methine group and therefore has an electron suction action on the electron gas vibrating in the conjugated chain. This electron suction is particularly conspicuous when the distribution of density at the site of substitution in the chain builds up in a certain place, since this state is assumed preferentially by the system for reasons of stability. The energy level in this case is lower than in a pure methine chain.

Thus absorption displacement depends on whether the place at which the energy accumulates and which is influenced by the substitution lies at the ground level or at the excitation level. If a maximum density coincides with a heterocyclic atom in the excitation term, then this state is formed preferentially and the energy of transfer ΔE, in the absence of any influencing of the ground state, decreases. Conversely, the ground level is lowered if the N atom meets a site of accumulation of energy in this state, and in this case ΔE increases.

Since the energy levels depend on the number of double bonds d and on the number of π-electrons, N, respectively, which are connected with the latter by the relation $N = 2d + 2$, the displacements in absorption which are observed on substituting the central C atom can be attributed to d or N, i.e. to the constitution of the dye. The following can be inserted in equation 23 for the centre of the conjugated system

$$\frac{s}{l \cdot N} = \tfrac{1}{2}, \tag{34}$$

so that the function of the ground state from which the electron density distribution $e|\psi_n^2|dv$ follows, can be simplified to:

$$\psi_{\text{centre}} \sim \sin\left[\frac{\pi}{2} \cdot n\right] (n = 1, 2, \ldots) \tag{35}$$

If the number of π-electrons N which is linked with the quantum number n of the energy state in accordance with $n = N/2$, or the number of double bonds d is inserted in equation 35, then the important relations for explaining the displacements in absorption will be as follows:

Ground state:

$$\psi_{\text{centre}} \sim \sin \frac{\pi}{2}\left(\frac{N}{2}\right) \tag{36a}$$

$$\psi_{\text{centre}} \sim \sin \frac{\pi}{2}(d + 1) \tag{36b}$$

Excitation state:

$$\psi_{\text{centre}} \sim \sin \frac{\pi}{2}\left(\frac{N}{2} + 1\right) \tag{37a}$$

$$\psi_{\text{centre}} \sim \sin \frac{\pi}{2}(d + 2) \tag{37b}$$

According to equation 36b the ground state has its maximum density in the centre of the system, and this gives rise to a hypsochromic displacement on substitution if the number of double bonds is an *even* number. As an example we would cite the azacyanines which possess an even number of double bonds between the ends of the conjugated chain and which exhibit hypsochromic displacements in absorption when the —CH= group in the centre of the compound is replaced by —N= [8][127,128,131].

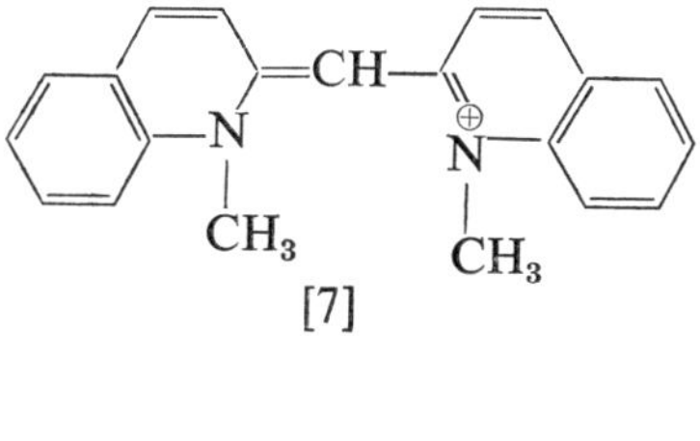

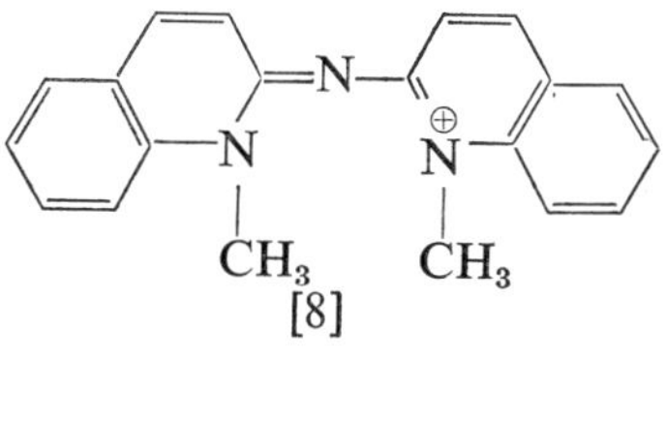

On the other hand, when a compound contains an *odd* number of double bonds, substitution produces a bathochromic displacement, since the ground state exhibits a minimum density at the site of substitution: nevertheless, according to equation 37b, the excited state shows a maximum. For the quantitative calculation of these displacement effects which are of the order of magnitude of 100 mμ, cf. Kuhn et al.[53,132,133]

The displacements in absorption are also influenced by the nature of the substituent attached to the central carbon atom in the methine chain. Whereas the replacement of the central hydrogen atom in benzothiatrimethine cyanine [9] by a phenyl group displaces the maximum absorption towards the longer wavelengths, methyl groups produce a hypsochromic effect.[134,135] The hypsochromic displacement may possibly be caused by the hyperconjugating effect of the methyl group, the σ-electrons of which act in a similar way to π-electrons. This assumption is substantiated by the fact that a meso-ethyl group brings about a smaller displacement.[136,137] On substituting the hydrogen atom attached to the central carbon atom by a heterocyclic nucleus, trinuclear cyanines are obtained: these compounds behave like asymmetrical cyanines (see sections 3.2.2 and 3.2.3 of this chapter).

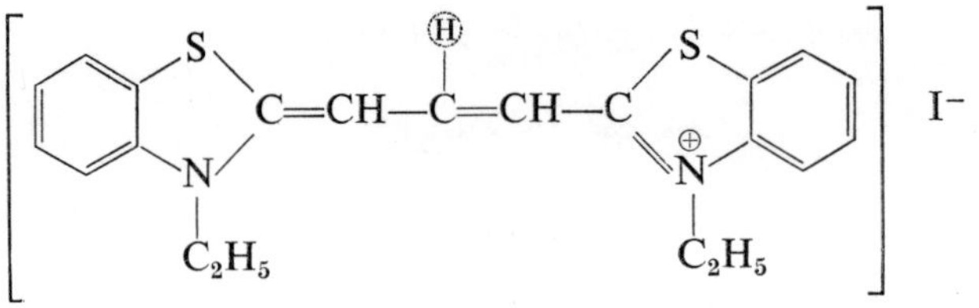

[9]

3. The displacements in absorption which occur when *substitution takes place in the nuclei* of dyes of the benzothiazole [10]-, benzoxazole [11]- or the quinoline series [12] are mainly bathochromic and are of the order of magnitude of several hundred Å units, being greatest in the 5- and 6-positions in the benzene ring.

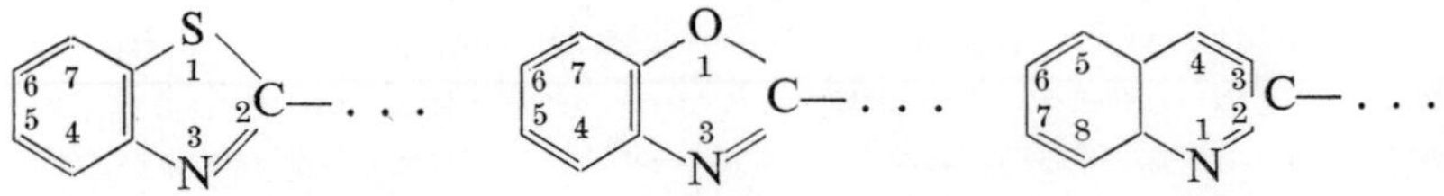

[10]   [11]   [12]

They increase in the sequence: methyl-, phenyl-, halogen-, methoxy-, methylthio-groups, benzene rings and dialkylamino groups,[126,134,138–141] in other words, in the order: —CH_3 < —C_6H_5 < —Cl < —OCH_3 < —SCH_3 < —NR_2.

4. The maximum absorption of *asymmetrical dyes* lies farther in the short wavelength region than that of the corresponding symmetrical compounds. This is evidently because the unsymmetrical terminal auxochrome and antiauxochrome groups, respectively, prevent the π-electrons in the conjugated chain from vibrating freely in the system.

It should be noted that auxochrome and antiauxochrome groups which have no intrinsic absorption in the visible region intensify the colour action. Auxochrome groups such as $-NH_2$, $-OH$, $-CH_3$, NR_2 endeavour to give up electrons to a conjugated system; antiauxochromes, such as $-NO_2$, $-COOH$, $-C\equiv N$, on the other hand, show a tendency to take up electrons.

By inserting these groups the basic structures are no longer so heavily involved in bringing about the mesomeric ground state, so that the degree to which the valencies are compensated decreases and the original single and double bonds reappear.[53,142] For this reason the potential between the C atoms can no longer be assumed to be approximately constant. The shape of the potential curve (see Fig. 10b) can only be adjusted by means of a sinusoidal variation in the potential energy, the amplitude of which, the so-called interference potential V_0 increases with the unsymmetry of the end groups and in the case of symmetrical end groups reaches values of from $V_0 = 0$ eV to $V_0 = 2$ eV.

The electron gas model takes into account the displacement in absorption brought about by lack of symmetry, by the introduction of the interference which is dependent on V_0, so that instead of equation 24 we have:

$$\Delta E = \frac{h^2}{8mL^2}(N+1) + V_0\left(1 - \frac{1}{N}\right). \tag{38}$$

In other words, an unsymmetrical system ($V_0 > 0$) absorbs at a shorter wavelength than does a symmetrical dye ($V_0 = 0$) of equal π-electron number:

$$\lambda_{max} = \frac{1}{\frac{h}{8mcl^2} \cdot \frac{N+1}{N^2} + \frac{V_0}{h \cdot c}\left(1 - \frac{1}{N}\right)} \tag{39}$$

The law governing asymmetrical dye-systems, which is known from experiments,[143,144] namely that after a specific wavelength the absorption maxima converge with increase in the number of π-electrons, can also be gathered from equation 39. This convergence is absent in symmetrical systems and λ_{max} increases in proportion to the number of π-electrons: for $V_0 \to 0$, and thus in this case equation 39 changes into equation 29.

It should also be noted that the interference potential V_0 necessary for calculating the absorption of asymmetrical systems can be derived from measurements of the long wavelength absorption bands of a dye in an asymmetrical series. To this end only λ_{max} and the allied π-electron number N have to be inserted in the interference equation 39.

In addition to this method, *a comparison of the basicity* of the heterocyclic nuclei attached to the ends of the polymethine chain enables the absorption maxima of asymmetrical dyes to be predicted with sufficient accuracy in practice,[144,160] because the position of the absorption bands is not determined solely by the symmetry of the heterocyclic nuclei, but also by the basicity and the electron donor action, respectively, of the latter.

In this connection it should be pointed out that according to the generalized acid-base concept of *Lewis*, bases can be regarded as molecules or ions which can give up pairs of electrons, whereas *Lewis*-acids are compounds which exhibit a tendency to accumulate electrons.

The influence of the basicity of the end groups can be seen from the fact that, of the symmetrical dyes of equal chain length, the compounds with the most strongly basic heterocyclic nuclei show the longest wavelength absorption. It is in these compounds that the mesomeric effect which leads to a stable electron configuration due to the interaction of the π-electrons—cf. for example, 53—is most sharply defined.

An example of a strongly basic nucleus is given by the pyridine ring in which the nitrogen atom actually assumes the role of an electron donor.

$$—N{\big<}\ \text{(donor)}\ \xrightarrow{-e}\ =\overset{+}{N}{\big<}\ \text{(acceptor)}.$$

By giving up electrons, the uncharged pyridine ring acquires an additional double bond, so that the molecule becomes stabilized aromatically as in the case of benzene. This system therefore has a great tendency to improve, i.e. to develop the basicity of the pyridine nucleus by giving up electrons.

If, say, the other end of a conjugated chain terminates in a positively charged, weakly acid pyridine nucleus, then, in the natural course of events, after having taken up electrons, the latter likewise exhibits strongly basic properties. The displacement of the electrons changes in direction, and therefore an uninterrupted and mutual exchange of electrons takes place between symmetrical end groups of this type.

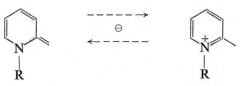

electron donor:　　　　　　　electron acceptor:
strongly basic　　　　　　　　weakly acid

A maximum exchange of electrons cannot be achieved by building heterocyclic rings of different basicities into the molecules. The mesomeric effect is less than in the case of rings of equal basicity, so that the absorption bands of asymmetrical dyes occur at shorter wavelengths than do those of symmetrical systems. These can be estimated on the basis of the differences in basicity of the heterocyclic nuclei with the aid of the following rules; cf. for example, *Zwicky* and *Brooker*.[134,144]

a. The maximum absorption of an asymmetrical dye with nuclei of similar basicity lies in the centre of the maxima of symmetrical dyes which are respectively composed of the same nuclei (*mean value rule*).

b. The maximum absorption of an asymmetrical dye with nuclei of strongly differing basicities no longer coincides with the mean arithmetical value of the maxima of the corresponding symmetrical dyes. Departures of $\Delta\lambda$ from this "mean value" of up to an order of magnitude of 100 mμ are possible. Here the absorption-maximum is situated in the vicinity of the maximum of a symmetrical dye absorbing in the short wavelengths, but approaches the latter on increasing the length of the polymethine chain, since the departures from the mean value increase on lengthening the polymethine chain: $\Delta\lambda = f(\text{number of methine groups})$.

An example of this is given by the asymmetrical cyanine dye [13] which is built up of a strongly basic pyridine nucleus and a weakly basic pyrrole ring, and the maximum absorption of which is situated at 426 mμ (in CH_3OH)[144,145]

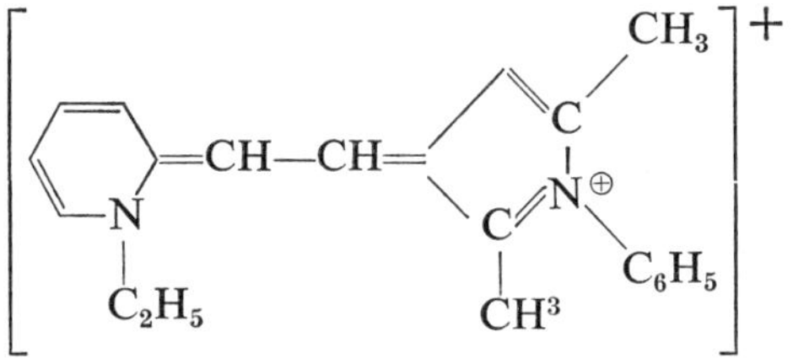

[13]

The symmetrical dyes [14, 15] which are based on the asymmetrical cyanine [13] and which are built up of 2,5-dimethyl-1-phenylpyrrole nuclei and pyridine nuclei, respectively, absorb at 449 mμ [14] and 562 mμ [15], respectively, in CH_3OH.

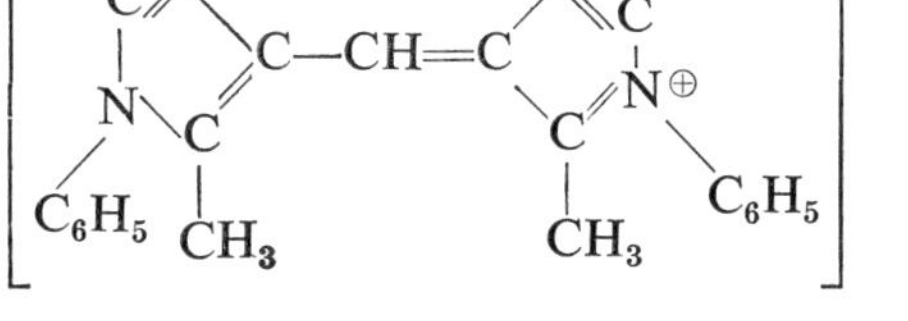

[14]

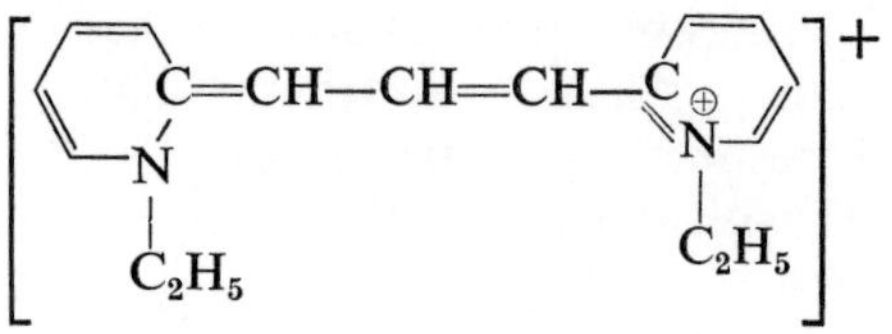

$$\left[\begin{array}{c} \text{pyridine-C=CH-CH=CH-C-pyridine} \\ \underset{C_2H_5}{N} \quad\quad\quad\quad\quad\quad \underset{C_2H_5}{\overset{\oplus}{N}} \end{array} \right]^+$$

[15]

c. The basicity of heterocyclic nuclei, and therefore also the maximum absorption of the cyanine dyes containing these nuclei can be altered by means of substituents.[145a] Nitro groups or halogen atoms diminish the basicity, whereas methoxy, ethoxy or methyl groups produce an increase. The displacements in the absorption $\Delta\lambda$ thus produced, as *Brooker*[144] has proved, are seen to depend on the Hammett-σ-constants of the substituents, so that the changes in absorption which are brought about in asymmetrical dyes by substituents, can, in certain circumstances, be derived from a $\Delta\lambda/\sigma$-relation. (For the definition of σ see section 2.2.6 of Chap. VI.

The influence of substituents may set a limit to the estimation of the maximum absorption of an unsubstituted asymmetrical dye by means of the mean value rule. This is particularly the case if the substituent produces such a large increase or decrease in the basicity of one heterocyclic ring, that the nuclei which are present in the asymmetrical dye are no longer of similar basicity.

For instance, whilst the absorption maxima of a dye consisting of a benzothiazole nucleus and a quinoline ring (attached in the 4-position to the polymethine chain) can readily be calculated by applying the mean value rule, this can no longer be done when a nitro group is introduced into the 6-position of the benzothiazole nucleus.[134] The p-nitro group causes such a large drop in the basicity of the benzothiazole nucleus as to produce a difference of $\Delta\lambda \approx 70\text{m}\mu$ from the mean value in the case of the trimethinecyanine [16].

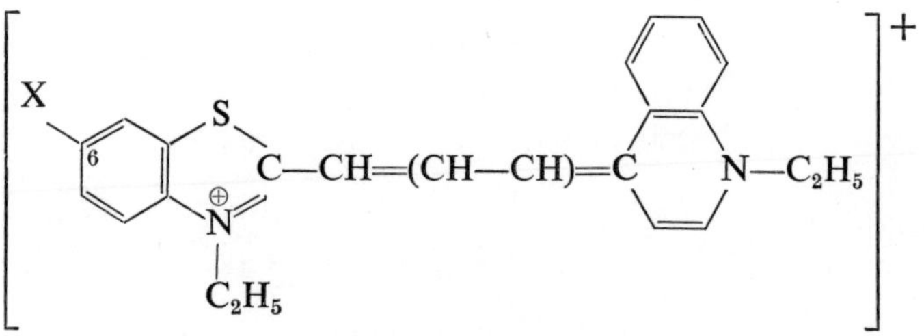

[16] $X = H: \lambda_{max} \approx 735 \text{ m}\mu$
$X = NO_2: \lambda_{max} \approx 665 \text{ m}\mu$

The method by which these effects are usually represented in the literature is shown in Fig. 12.

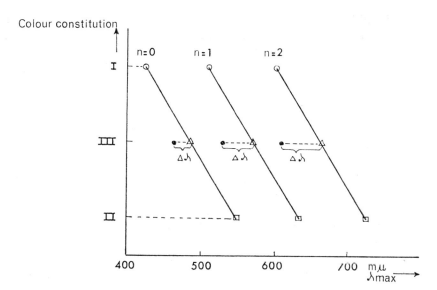

Fig. 12. Diagrammatic representation of the displacements in the absorption of asymmetrical dyes. *Explanation of the symbols:* I, II: Standard symmetrical dyes from which an asymmetrical dye is synthesized. Combination in the case of I from weaker basic nuclei than in the case of II. III: Asymmetrical dye: Δ: λ_{max} when the mean value rule is valid. $\bigcirc$: λ_{max} when the mean value rule is not valid owing to strong variation in the basicity of the I and II nuclei. Here λ_{max} converges with increase in n.
Note: $\Delta\lambda = f$ (Hammett-σ).

d. According to *Brooker*,[144] a slight chemical change has less effect on the absorption behaviour of symmetrical than of unsymmetrical dyes. For example, the substitution of the benzothiazole nucleus by the more strongly basic 2-quinoline nucleus in a symmetrical benzothiacarbocyanine only displaces the absorption of the latter by the negligible amount of 3 mμ. If, however, the benzothiazole nucleus in an asymmetrical benzothiazole-pyrrole methine dye is replaced by 2-quinoline, this results in an appreciable displacement in absorption of about 15 mμ.

In summarizing the situation it can be stated that it is possible to discover an organic dye of suitable absorption for every wavelength in the visible and near infra-red regions of the spectrum. Here, however, the question arises as to how far the dyes absorbing in the various spectral regions also have sensitizing properties and what relations exist between dye constitution and spectral sensitizing power.

2.2 *The Connection between the Constitution of a Dye and its Sensitizing Action*

2.2.1 THE EFFICIENCY OF SENSITIZATION. A successful method of testing the suitability of dyes for sensitization is the sensitometric method described in

Chap. I. In addition to the change in the spectral sensitivity of an unsensitized photographic material brought about by the sensitizer, the order of magnitude of the relative quantum yield Φ_r is of particular interest. This is because Φ_r is a measure of the efficiency of the transfer of the light-energy taken up by the dye, to the primary photographic material—also including electrophotographic sensitive layers—and indicates whether a dye is a good or a bad sensitizer for a specific region of the spectrum.

The definition and measurement of Φ_r have already been discussed in section 2.6 of Chap. I. Here we shall only supplement this by calling attention to the processes which can influence Φ_r and therefore also the efficiency of the transfer of energy and the sensitization, respectively. In theory, the sensitizing action can be diminished or entirely suppressed by the deactivation of dye molecules which are in a state of singlet excitation, after having absorbed light, as follows.

1. By deactivation by radiation in the form of fluorescent light (k_f).
2. By radiationless internal deactivation (k_i).
3. By radiationless external deactivation, which, e.g., occurs on changing over to photographically inert acceptors (antisensitizers, etc.) (k_a).
4. By intersystem-crossing to a triplet state which may not be used for the photographic process,[146,147] and from which point onwards deactivation by radiation or a photochemical process (bleaching-out process) sets in (k_t).

If the individual deactivating processes which are directed against the sensitization effect are denoted by the coefficients k_f, etc., which are inversely proportional to the average lifetime of the singlet excitation state which persists to the end of the process taking place at the time, then the quantum yield can be estimated from equation 40.

$$\Phi_r = \frac{k_s}{k_s + k_f + k_i + k_a + k_t} \qquad (40)$$

Since $b = 1/\Phi_r$ can be established for the quantum requirement

$$b = 1 + \frac{k_f + k_i + k_a + k_t}{k_s}, \qquad (41)$$

it is understandable that it is only in the case of negligible deactivation processes ($k_f + k_i + k_a + k_t \ll 1$) that the energy corresponding to an absorbed quantum can be transferred to the photographic material.

Deactivation by fluorescence has no influence in the case of adsorbed dyes ($k_f \to 0$), and when electron acceptors are absent k_a can also be neglected. If that part of the energy which has passed into the triplet state, and which in certain circumstances is even capable of promoting the sensitization process, is disregarded, then it can be seen that, in the main, the quantum yield is diminished by radiationless deactivation processes.

This does not, however, rule out the possibility that in many cases a small quantum yield Φ_r may be the result of some internal deactivation which is dependent on the configuration of the sensitizer.

West and *Carroll*[38] attribute, for instance, the poor or absent sensitizing power of dyes of irregular constitution to the presence of low frequency torsion vibrations which are excited during radiationless deactivation. In harmony with this is the fact that these dyes do not fluoresce in solution, coupled with their quenching action on the fluorescence of planar compounds. Even the decrease in Φ_r in compounds containing relatively long polymethine chains would be in agreement with this argument, since long chains exhibit a greater tendency to vibrate.[148,149]

However, the extent to which the internal deactivation of the excitation energy can generally be assumed to be the actual cause of the poor sensitizing power of a dye remains uncertain. A planar dye does in fact likewise lose its sensitizing power if, for example, it is not adsorbed directly to a silver halide grain. In this case radiationless internal deactivation is not the cause, but the *consequence*, of the absence of sensitization.

In order to explain the intensity of sensitization which is interpreted by means of the relative quantum yield, the co-operation of several effects respectively, that is to say, the mechanism of the sensitization process as a whole must be considered.

2.2.2 GENERAL PREREQUISITES FOR THE SUITABILITY OF DYES FOR USE AS SENSITIZERS. For a dye to be suitable for use in the production of spectrally sensitized emulsion layers, it must conform with a number of prerequisites which are not directly connected with the actual process of sensitization. These provisions are often also adapted to special emulsion problems.[134,150]

Firstly, any action on the part of the dye which is tantamount to an increase in the fog of an unexposed photographic emulsion layer should be as slight as possible. Whereas most of the sensitizers which are used in practice fulfil this condition satisfactorily, dyes with two quinoline nuclei (Pinacyanol), or with dialkylamino groups, and particularly infra-red sensitizers, are noted for their appreciable fogging action.[60]

Secondly, after a sensitive layer has been processed, the dye should be easily removable by washing in order to remove any colour which may be left behind in the photographic or electrophotographic materials. Solubility in water is the most important factor in the washing process, but a role is also played by the geometrical configuration of the molecules, since dyes of spherical shape diffuse out of an emulsion more readily than do elongated ones.

Solubility is also important for the introduction of a dye into an emulsion. The sensitizer should indeed envelop the silver halide grains to the fullest possible extent. When its solubility is low, large volumes of solution naturally have to be used for sensitizing a material, and in certain conditions this has a deleterious effect on the photographic properties of the emulsions or of the semiconductor layers (ZnO, etc.). In the case of cyanine iodides which are often sparingly soluble, this difficulty can be overcome by changing over to the more soluble bromides and chlorides. The water-solubility of these compounds can also be improved by the introduction of special groups which include N-hydroxyalkyl-, N-carboxyalkyl-, $H_2NSO_2(CH_2)_3$-, $CH_3SO_2NHCOCH_2$-, $(CH_3)_3N^+$—$(CH_2)_3$- groups, etc.[144, 151–154]

In the manufacture of colour film the opposite effect is often striven for, the aim being to produce dyes of low diffusing power in order to prevent non-adsorbed sensitizers from diffusing between the emulsion layers in the colour film.[155] With colour emulsions it is also important for favourable conditions to exist for the adsorption of the sensitizer, even when fairly high concentrations of added coupler molecules are present in the emulsion. First and foremost, combination between the sensitizer and the coupler to form a salt must be avoided as this causes the sensitizer molecules which have already been adsorbed by the silver grains to become desorbed and dislodged by the coupler molecules.[64, 90] This trouble is removed by introducing a carbonic acid, a sulphonic acid or a sulphate group into the alkyl- or aryl-residue of the nitrogen atom in the heterocyclic ring of sensitizers.

In analogy to the last-named effect, the intensity of sensitization is also greatly influenced by any interaction which may take place between the sensitizer molecules and coating additions such as stabilizers, wetting agents, gelatin hardening agents, etc. The sensitizing effect decreases if the dyes are displaced by the substances added to the emulsion. An essential prerequisite for the sensitizing power of a dye is therefore that it must have a strong tendency to be adsorbed to the surface of a grain. Since emulsions can differ from one another because of the different substances added to them, in certain circumstances anomalous displacement effects lead to variations in the intensity with which a dye sensitizes different emulsions.

2.2.3 THE EFFECTS PRODUCED BY SUBSTITUTION. The bathochromic or hypsochromic displacements in absorption brought about by substitution in the conjugated chain of a dye (see section 1.3/2 of this chapter) are connected with the sensitizing power of the dye.[156] *Riester*,[130] by comparing the spectral and photographic behaviour of various dyes, was able to show that in the case of pairs of dyes of similar constitution, the dye which, after substitution, absorbed at a longer wavelength, often had desensitizing properties, whereas the non-substituted compound which absorbed at a short wavelength, sensitized. This regularity holds not only for the oldest sensitizers such as Eosine, but also for cationic cyanine dyes or neutral merocyanines.

As an example of this we would cite Eosine [17] and Iris blue [18] which is obtained by substituting the bridging carbon atom by nitrogen. Whereas Eosine has a maximum absorption at 525 mμ and sensitizes, Iris blue, which absorbs at 603 mμ, is a desensitizer. Analogously, Methylene red [19] with a $\lambda_{max} = 565$ mμ acts as a sensitizer, whereas Methylene blue [20] with a $\lambda_{max} = 667$ mμ has a desensitizing action.

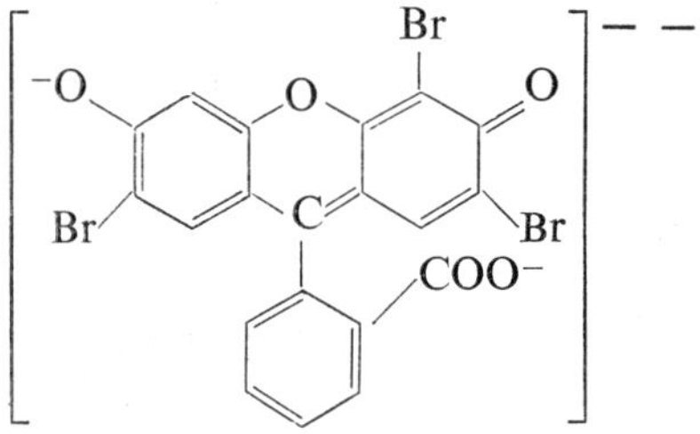

[17]

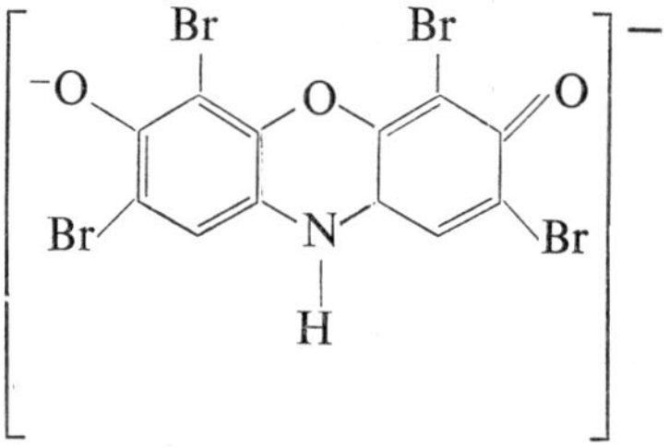

[18]

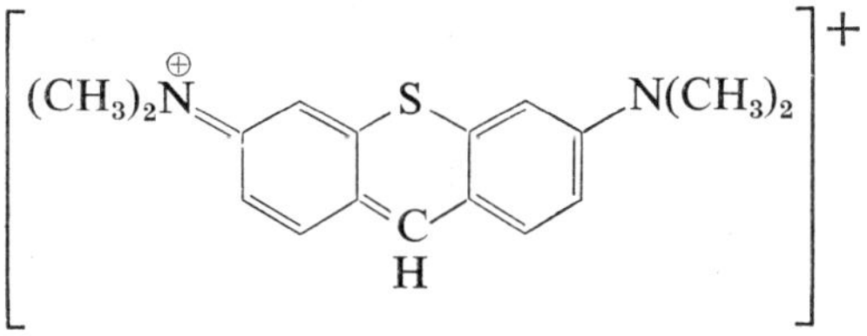

[19]

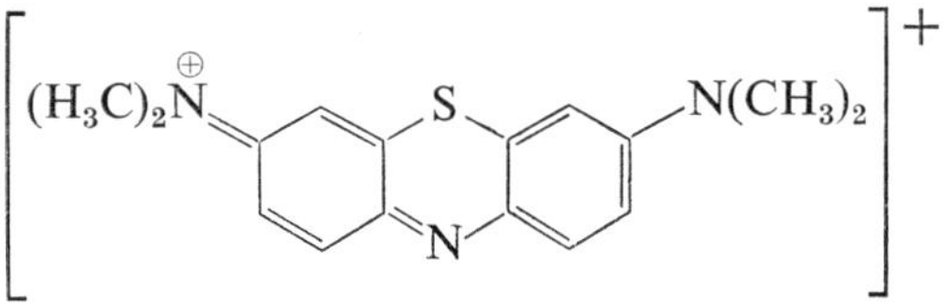

[20]

The same effect is produced by substituting a methine group in the polymethine chains of cyanines, by a nitrogen atom. For instance, the Aza isologue [22] of the sensitizer Astraphloxine [21] ($\lambda_{max} = 544$ mμ) absorbs at a longer wavelength ($\lambda_{max} = 595$ mμ) and behaves as a desensitizer.

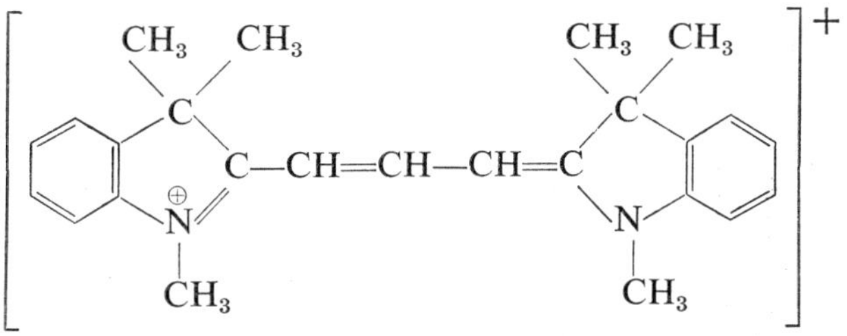

[21]

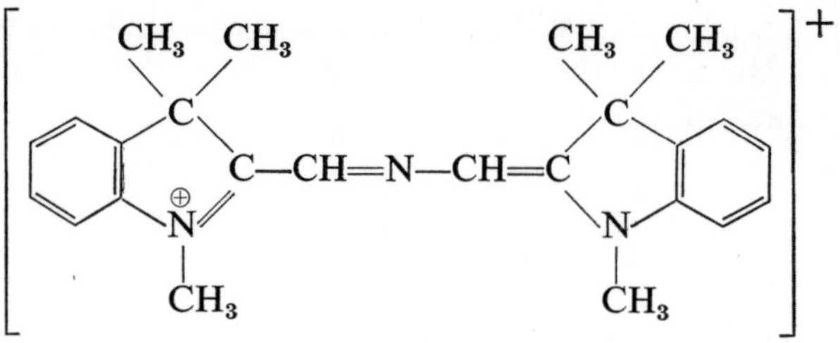

[Structure [22]: bis-indoleninium azacyanine dye]

[22]

This connection between the constitution and sensitizing power of dyes is not limited to the substitution of the =CH— group by =N—. The introduction of a nitrile group and other substituents enables analogous regularities in behaviour to be recognized. Moreover, the bathochromic or hypsochromic displacement in absorption in relation to the absorption of the parent dye is also dependent on the sensitizing or desensitizing effect of the position of the substituents in the polymethine chain.

If, for instance, Astraphloxine [21] (λ_{max} = 544 mμ) is substituted in the meso- or the β-position, respectively, by a cyanogen group, a dye which is analogous to Azacyanine [22] and which desensitizes and absorbs at 600 mμ is produced. However, the introduction of a nitrile group into the α-position displaces the maximum absorption towards shorter wavelengths and gives the dye a sensitizing action.

The example given by *Riester*[130] of Thiazol purple substituted in this way in the α- and β-positions respectively, also enables this regularity in behaviour to be seen clearly. The parent dye [23] is a sensitizer which has a maximum absorption at 559 mμ. The β-cyanogen substituted dye [24] exhibits an absorption band which is bathochromic to [23] (λ_{max} = 610 mμ) and desensitizes, whereas the β-cyanogen substituted dye [25] absorbs at λ_{max} = 520 mμ and again sensitizes.

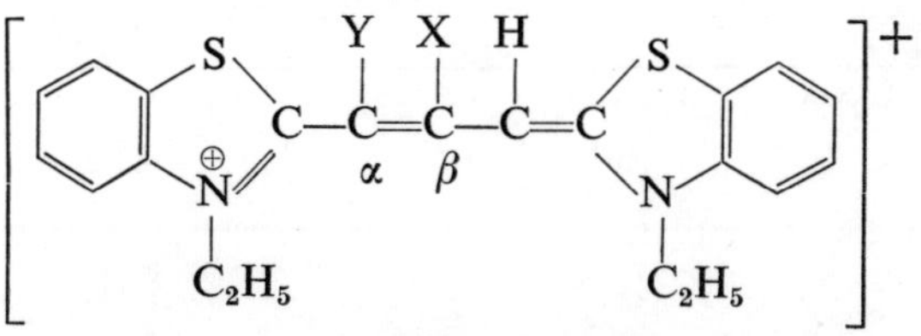

[23–25]

Thiazol purple [23] X, Y: H; λ_{max} = 559 mμ; sensitizer.
Dye [24] Y: H; X: —CN; λ_{max} = 610 mμ; desensitizer.
Dye [25] Y: —CN; X: H; λ_{max} = 520 mμ; sensitizer.

Similar regularities which are connected with the displacement in absorption on substitution and the reversal of the photographic behaviour can be observed with many dyes. Here we shall only mention Pinaflavol [26], which loses its sensitizing action on substituting the β-methine group by nitrogen, at the same time undergoing a bathochromic displacement in absorption, and desensitizing. If, however, nitrogen is introduced instead of the α-methine group, a sensitizer which absorbs in the short wavelengths is obtained.

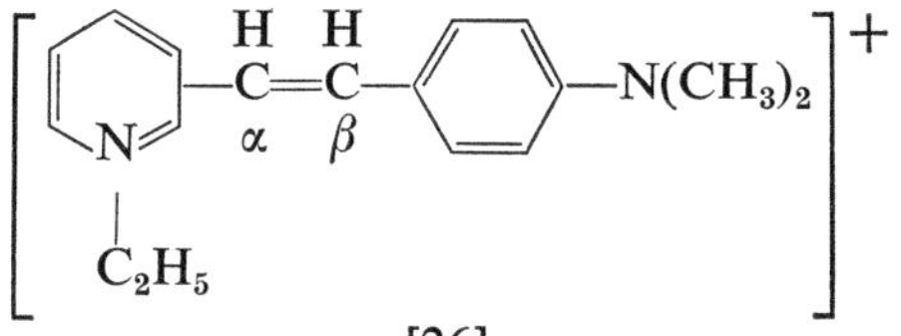

$$\left[\underset{\underset{C_2H_5}{|}}{\underset{N}{\bigcirc}}-\underset{\alpha}{\overset{H}{\underset{|}{C}}}=\underset{\beta}{\overset{H}{\underset{|}{C}}}-\bigcirc-N(CH_3)_2\right]^+$$

[26]

Even the substitution of the central H atom by a methyl group effects changes in absorption and sensitization. For example, mesomethylbenzothiatrimethine cyanine [27] which has a shorter wavelength absorption than that of benzothiatrimethine cyanine is a more efficient sensitizer. Substitution with larger aliphatic residues also has a similar, even if less powerful, effect.

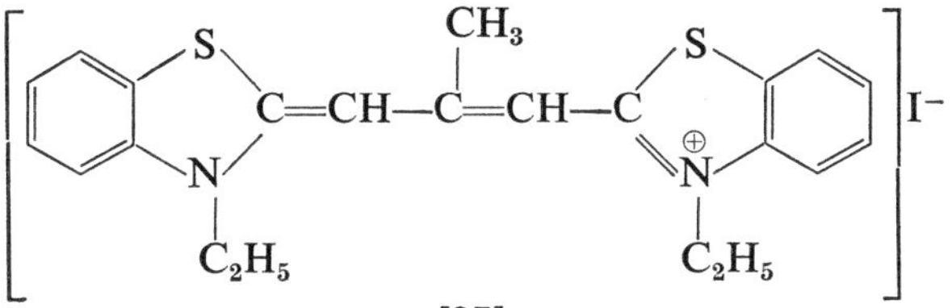

$$\left[\underset{\underset{C_2H_5}{|}}{\overset{S}{\underset{N}{\bigcirc}}}C=CH-\overset{\overset{CH_3}{|}}{C}=CH-\underset{\underset{C_2H_5}{|}}{\overset{S}{\underset{\overset{\oplus}{N}}{\bigcirc}}}C\right]I^-$$

[27]

For further examples, cf. 130, 144.

These regularities in behaviour often enable the photographic properties of a dye to be arrived at from its constitution, or the behaviour of the said dye to be explained. Taking diazocarbocyanine [28][157] as an example, this dye will

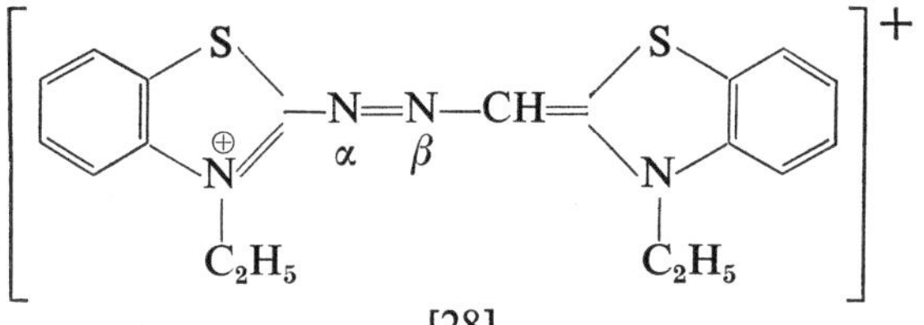

$$\left[\underset{\underset{C_2H_5}{|}}{\overset{S}{\underset{\overset{\oplus}{N}}{\bigcirc}}}-\underset{\alpha}{N}=\underset{\beta}{N}-CH=\underset{\underset{C_2H_5}{|}}{\overset{S}{\underset{N}{\bigcirc}}}\right]^+$$

[28]

obviously act as a desensitizer owing to the β-position of the nitrogen atom. The opposite effect of the α-nitrogen atom is presumably masked to a great extent, but is indicated in the weak sensitizing property of the dye in the green region of the spectrum.[144]

The relationship between the constitution and absorption of a dye, and its photographic behaviour, respectively, as illustrated by these examples, makes it natural to assume that it may be possible to describe the sensitizing power, in analogy to light-absorption, by means of the electron gas model, since the connection between the bathochromic effect and desensitization shows that the fall in the excitation level brought about by the substituent must be linked in some way with the decrease in sensitizing power. *Scheibe* and *Dörr*[158] regard the change in the position of the excitation level due to this effect as the actual cause of the reversal from sensitization to desensitization. How far this shift in the level suffices on its own to explain spectral sensitization and desensitization, respectively, will not be discussed any further here. The main question to be answered is whether the change in absorption possibly represents only a side phenomenon of the effect of constitution which gives rise to sensitization and desensitization, respectively.

The experimental findings might, however, be taken to prove that the sensitizing power of dyes is decreased by electron-attracting substituents meeting the places in which electrons have accumulated in the conjugation system which is in a state of excitation. However, if the electron-acceptor coincides with an electron density maximum of the ground state, which maximum can be determined from equation 23, then in certain circumstances, the substituted system will exhibit a better sensitizing action than the non-substituted system. According to equations 23, or 36 and 37 respectively, these regularities in behaviour are closely connected with the number of π-electrons, the number of carbon and nitrogen atoms, and the number of double bonds, respectively, since the electron density distribution and the occupation of the levels naturally depend on these constitutional properties.

Basically the same state of affairs is expressed in the rules formulated by *Kendall* [56] and extended by *Riester*[130] to explain the relation between constitution and sensitization and desensitization. For the *Kendall-Riester* rule to the effect that a substituent which exerts a negative influence and which is attached to a positively induced member in the chain, causes the substituted dye to absorb at a longer wavelength than the unsubstituted dye, and to desensitize, can ultimately be reduced to the characteristic electron density distribution along the conjugated chain. This is also true of the rule that the same substituent attached to a negatively induced member in the chain, causes the absorption maximum to be pushed back, and, in certain circumstances, the sensitizing properties of the parent dye to be re-formed.

The nature of the positive and negative induction produced by auxochrome groups in the conjugated members in the chain will be illustrated, taking a trimethinecyanine dye [29] as example. Here it is essential for the auxochromes A, A' to act on the methine groups as inducing key atoms. Undoubtedly in these formulations we are dealing with extreme forms. The electron distribution is, however, given in suitable approximation to enable the displacement effects and the photographic properties to be predicted.

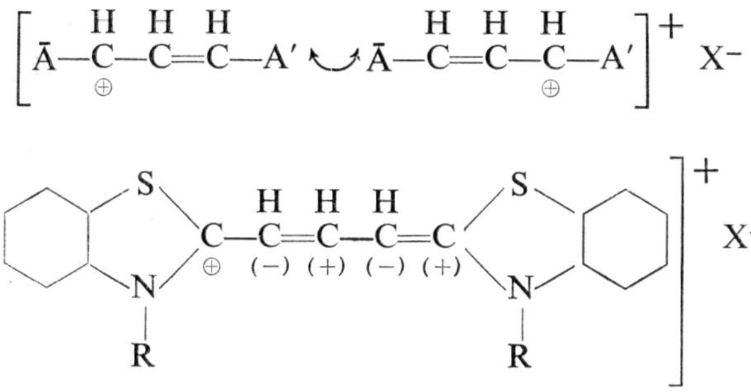

$$\left[\overset{\text{H}}{\underset{\oplus}{\text{A}-\text{C}}}-\overset{\text{H}}{\text{C}}=\overset{\text{H}}{\text{C}}-\text{A}' \longleftrightarrow \overset{\text{H}}{\text{A}}-\overset{\text{H}}{\underset{\oplus}{\text{C}}}=\overset{\text{H}}{\text{C}}-\text{C}-\text{A}' \right]^+ \text{X}^-$$

[29]

2.2.4 THE INFLUENCE OF THE SPATIAL STRUCTURE OF A DYE ON ITS PHOTOGRAPHIC BEHAVIOUR. The photographic behaviour of a dye is decided not only by its chemical constitution, but also by the spatial configuration of the dye molecules. This was first recognized by *Sheppard*[159] in the year 1941 and was subsequently substantiated by *Brooker et al.*[160,161] by means of numerous examples. The experimental material now available provides the basis for assuming that the *planarity* of a dye is undoubtedly the most important geometrical prerequisite for its sensitizing power.

In the case of the cyanines this signifies that for effective sensitization to take place the polymethine chain with its two heterocyclic nuclei must lie in one plane. As soon as a dye with a flat construction loses its planarity, its sensitizing power decreases greatly and may even be entirely lost. Thus the sensitization effect is very susceptible to changes in the flat configuration of the dye. For instance, the sensitizing power can even be diminished by the introduction of substituents which have hardly any effect on absorption.

Here we shall only mention the removal of the sensitizing power of Pinacyanol [30] when the central H atom in the methine chain is substituted by a methyl group [31].

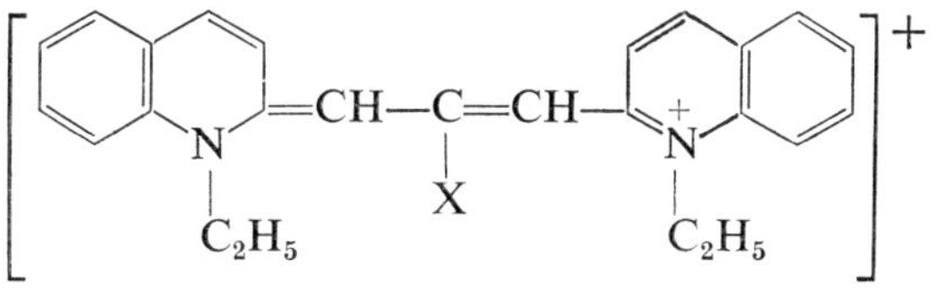

[30] X : —H Pinacyanol, sensitizer;
[31] X : —CH$_3$ no sensitizing power.

According to the three-dimensional representation shown in Fig. 13, once methyl groups have been built into the molecule, the latter ceases to have a flat configuration, so that its sensitizing power must regress in accordance with the planarity hypothesis.

This classification of dyes into planar and non-planar groups, which is of great importance from the point of view of sensitization, has been substantially extended by *Brooker*.[161]. *Brooker* distinguished the following groups of dyes according to the extent to which their components occupied all the space in the molecules.

1. The group containing the loosely packed planar dye molecules, the charac-

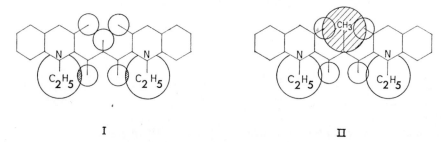

Fig. 13. Examples of a flat sensitizing dye (I) and of a non-flat dye (II).

teristic of which is that there is relatively more free space between the non-combined atoms. This group contains, for example, the benzoxazoletrimethine cyanine presented three-dimensionally in Fig. 14.

2. The group of the compact planar molecules in which all the free space is entirely filled up owing to the increase in the number of atoms or groups which

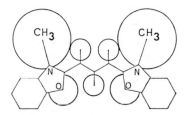

Fig. 14. Example of a loosely-packed planar dye.

are built into the molecule. The most powerful sensitizing dyes are to be found in this class.

3. The groups of the overlapping or crowded dye molecules which do not have a flat configuration and therefore in general no longer have any sensitizing action. There are, however, exceptions within this group of dyes some of which have a weak sensitizing power and claim in particular to be supersensitizers.[144]

A further rule of special interest in connection with the planarity hypothesis is as follows. Whereas overlapping, non-planar dyes which have no sensitizing action are produced by the introduction of methyl groups, etc.—as in the case of Pinacyanol [30]—the introduction of larger nuclei into the meso position in cyanines does not often have an analogous inhibiting action on their sensitizing properties. In certain circumstances it appears that such multinuclear dyes react like compact sensitizers although there is not sufficient space to accommodate the nuclei in one plane. A dye of the neocyanine type [32] is quoted as an example. This dye is a good sensitizer.

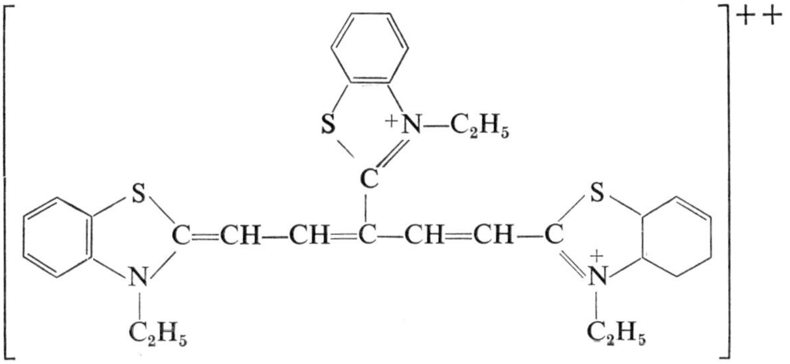

[32]

An explanation of the apparent contradiction to the *Sheppard* and *Brooker* planarity rule has been given by *Bruylants*, *van Dormael* and *Nys*[162] according to whom the planarity of the chromophore system in cyanine dyes can be assumed to be maintained if the substituent is rotated out of the plane of the other two nuclei. The fact that the additional nucleus no longer participates in the resonance system is shown by the small influence of this nucleus on the absorption. Cf. also 144, 150, as well as section 3.2.3 of this chapter.

The spatial configuration of sensitizers has the additional effect of enabling compact cyanine dyes to form *J*-aggregates more readily than is possible with dyes of less compact structure. This rule also holds for those multinuclear dyes which have retained a compact chromophore system by the outward rotation of a nucleus. In certain circumstances loosely packed dyes can likewise be made to form *J*-aggregates by the introduction of atoms if the latter promote the mutual attraction of the flat molecules, and therefore the aggregation. This is observed for example, in the loosely packed pinacyanol molecule when chlorine is introduced in the 6,6'-position. However, it must also be remembered that in vinylhomologous series, the ability to form *J*-bands usually diminishes with increase in the length of the methine chain.[149]

In spite of the comprehensive material available on the relation between the spatial structure of molecules and their sensitizing power, the cause of this relation has not yet been elucidated. The difference in the behaviour of the individual groups of dyes undoubtedly suggests that the excitation energy can be more easily deactivated by transformation into internal vibration energy in more loosely packed than in rigid, compact molecules. How far this is the actual cause of sensitization which is sensitive to any departure from planarity, is still uncertain.

In summarizing it may be said that the 3rd and 4th sections do, however, reveal that the introduction of substituents influences the sensitizing power of dyes both by disturbing the electronic vibration system as well as by means of spatial effects.

2.2.5 THE EFFECT OF AGGREGATION. It has already been pointed out in section 4 that particularly compact dyes often form characteristic *J*-aggregates. This formation of *J*-aggregates which is revealed by the sharper absorption at longer wavelengths than in the case of the monomer, influences not only the position of the sensitization band but also other aspects of the effect produced by the sensitizing dye.

For instance, the formation of *J*-aggregates renders the spectral sensitivity[163] and the contrast of sensitized emulsion layers appreciably dependent on the concentration of the dye. In other words, whilst at low concentrations aggregation of the adsorbed dye is only established at particularly suitable grain surfaces, at high concentrations practically all the adsorbed molecules are already present in the *J*-state. Thus, at low sensitizer concentrations an emulsion which is sensitive in the short wavelength regions corresponding to the monomer is insensitive in the spectral region of the *J*-band. It is only from a minimum concentration onwards that sensitivity occurs in the region of the *J*-band, and this sensitivity increases with increase in the concentration of the sensitizer.

The gamma value (contrast) of the *J*-region differs from that of the monomeric region in which it is largely constant, in being very small at low concentrations, and it is only at higher dye concentrations that it increases to a limiting value; cf. in addition *West* and *Carroll*.[38]

It should be noted that the effects due to the *J*-associates are strongly influenced by the type of emulsion[162a] and by the surface of the silver halide. The formation of associates often depends entirely on the adsorption to the surface of the silver halide. This is apparently consequent on a straightening out of the dye molecules by the crystal lattice, and it is for this reason that *J*-aggregates form at lower concentrations on well developed crystal surfaces than on poorly developed ones.[111]

Moreover, the (100) surfaces differ from the (111) surfaces of silver halide crystals in their ability to become sensitized. The *J*- and *H*-bands occur more readily at the less distorted (100) surfaces of cubic crystals than at the surfaces of octahedral crystals, which latter are more sensitive to microcrystalline lattice defects. This difference, which is discernible in ammoniacal emulsions and in boiled emulsions, has an effect not only on the absorption behaviour of the adsorbed dye,[163a] but also on the shape of the sensitization curve. According to *Markocki*,[164] silver bromide emulsions containing mostly cubic crystals show up to 50 times greater sensitivity in the region of sensitization than do emulsion layers which are built up mainly of octahedral crystals such as those which can be obtained by controlled Ostwald ripening.[165] On the other hand, the desensitizing action is particularly marked when the dyes are adsorbed to the (111) surfaces.

The difference in the behaviour of cubic and octahedral crystals of silver halide is also of practical importance, but how far lattice distortions alone are responsible for this difference—cf. in addition 110—remains uncertain.[111,166] The geometrical form of the crystals may possibly have a decisive influence on the adsorption and aggregation of large and rigid dyes. Investigations of the adsorption of dyes,[167] particularly to well-defined crystal surfaces such as those described by *Günther* and *Moisar et al.*[168,168a] appear to be of great interest in clarifying this question.

2.3. *The Sensitizing Dyes used in Practice*

The relations between dye constitution and sensitization which have been discussed represent the basis of the development of the modern spectral sensitizing dyes which are nowadays available in large numbers to the emulsion-maker.

Sensitizers belong practically without exception to the polymethine class of dyes,[126,169] and whether they can be used in practice largely depends—as can be seen from sections 2.1 and 2.2—on the elements of their structure: namely

1. On the length of the polymethine chain which is built up of an odd number of conjugated carbon atoms.
2. On the nature of the electron-acceptor- and donor-groups combined by means of the polymethine chain.
3. On the substituents which may be present in the chain or attached to the end groups.
4. On the extent to which the space in the molecule is occupied.

Since the questions which are connected with the above conditions have already been discussed at some length, the systematics of the sensitizers used will be dealt with only briefly in the following section. For details, cf. the reviews written by *Zwicky, Riester, Hamer, Brooker et al.*[64,126,134,144,150,169,170]

Theoretically, the large number of sensitizers which are in common use nowadays fall into two large classes: dyes with two nuclei which belong to the fundamental type of polymethines and sensitizers containing several nuclei which number among the polymethine dyes of higher order.

2.3.1 THE FUNDAMENTAL TYPE OF POLYMETHINE DYE. The molecules obtained by attaching each of the two ends of a chromophore polymethine chain to auxochromic heterocyclic atoms (nitrogen, oxygen, phosphorus) which usually, together with the neighbouring carbon atoms, are usually built into ring systems, bear different charges, depending on the nature of the heterocyclic atoms. This charge forms the basis of the division of polymethine dyes into cationic, anionic and neutral dyes.

2.3.1.1 *Cationic or Basic Polymethine Dyes.*

2.3.1.1.1 Structure. In this important group, both ends of the chromophore system terminate in nitrogen atoms which in the extreme case carry the positive charge of the dye molecule in accordance with the following general formulae for representing mesomers

$$\left[>\!N\!-\!\overset{|}{C}\!=\!\left(\overset{|}{C}\!-\!\overset{|}{C}\!=\!\right)_n\!\overset{\oplus}{N}\!<\right]^{+} \rightleftarrows \left[>\!\overset{\oplus}{N}\!=\!\overset{|}{C}\!-\!\left(\overset{|}{C}\!=\!\overset{|}{C}\!-\!\right)_n\!N\!<\right]^{+}$$

This charge is compensated by some kind of anion which, although it does not influence the absorption of light, is however able to influence the solubility of the dye in the sequence $F^- > HSO_4^- > NO_3^- > Br^- > SCN^- > I^- > ClO_4^-$.

The cationic group includes above all the important sensitizers, the cyanines themselves. Whereas in the latter both of the nitrogen atoms are built into the rings, which at the same time contain a part of the polymethine chain, this is not the case with the streptocyanines [33] and the hemicyanines [34] which are of interest in the synthesis of cyanines of long chain length, and of unsymmetrical cyanines, respectively. It should be noted that di- and triphenylmethane dyes may be regarded as phenylogues of the streptocyanines, and styryl dyes—which include for example, Pinaflavol [26]—as vinylogues and phenylogues of the hemicyanines.

$$\left[\begin{array}{c}R\\ \!\!\!\setminus\!\!\overset{+}{N}\!\!=\!\!C\!-\!\!\left(\!\!\begin{array}{c}H\ H\\ C\!=\!C\end{array}\!\!\right)_{\!n}\!\!-\!\!N\!\!\begin{array}{c}R\\ \!\!\diagdown\!\!\\ R'\end{array}\\ R\end{array}\right]^{+} \qquad \left[\begin{array}{c}Z\!\!\diagdown\!\!\\ \!\!\!\!C\!-\!\!\left(\!\!\begin{array}{c}H\ H\\ C\!=\!C\end{array}\!\!\right)_{\!n}\!\!-\!\!N\!\!\begin{array}{c}R\\ \diagdown\\ R\end{array}\\ \overset{+}{N}\!\!\diagup\\ |\\ R\end{array}\right]^{+}$$

[33]   [34]

Numerous nuclei are used as heterocyclic ring systems in cyanine dyes, and a few of these are compiled in Table 5; cf. in addition the review on the subject by *Zwicky*.[150]

According to the rules discussed in section 2.1 of this chapter, the positions of the absorption maxima are greatly dependent on the heterocyclic bases which are built into the dyes. It is, moreover, of particular importance whether equally basic nuclei (symmetrical cyanines) or different heterocyclic nuclei (asymmetrical cyanines) participate in the structure of the polymethine dyes. In addition, a number of characteristic possibilities of influencing the sensitizing properties are offered by varying the nature of the *substitution*.

TABLE 5
BASIC RING SYSTEMS OF THE CYANINES

System	Structure	System	Structure
Pyrrole	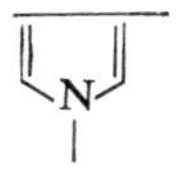	Pyrazolenine	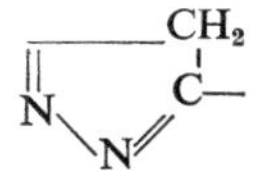
Oxazole	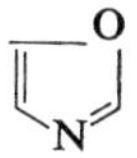	Thiazole	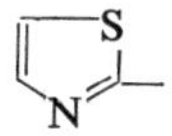
Pyridine	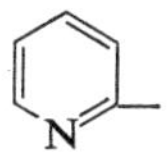	Pyrimidine	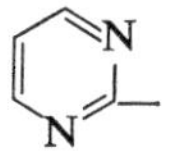
Benzimidazole	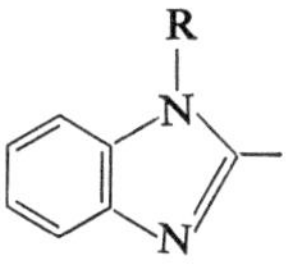	Benzothiazole	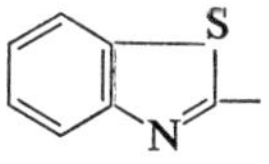
Indolenine	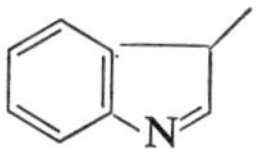	Quinoline	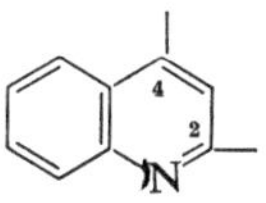
β-Naphthothiazole	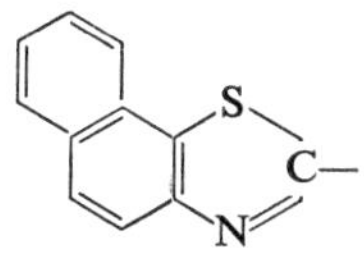	α-Naphthothiazole	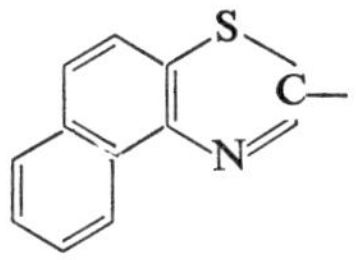
Thiazoline	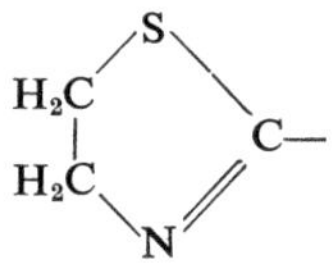	4-Phenylthiazole	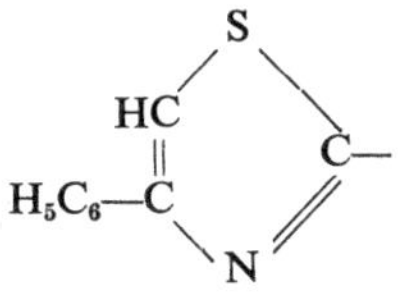

1. Substitution can influence absorption behaviour by the changes which it produces in the basicity of the *original rings*, and which often also improve the sensitizing property of a dye. In addition to alkyl and aryl groups, the principle substances used for substitution in the ring are condensed benzene rings, halogens, alkoxy and thioether groups, cyclopentane rings, dialkylamino groups, acyl groups, nitrile groups and so on. The 5- and/or the 6-positions in the benzene ring are the optimum ones.

2. Substituents attached to the *ring nitrogen atom* have hardly any effect on the position of the absorption band, but the suitability or otherwise of a

dye for use in sensitization may be greatly dependent on this substitution. For instance the spectral sensitivity of N,N'-ethyl-meso-ethyl-thiacarbocyanines in the region of sensitization is raised by the introduction of a hydrophilic carboxyl group. Esterification of the ethyl-carboxyl group, on the other hand, brings about hydrophobization of the molecule, and this leads to a sharp fall in sensitivity.[171] Whereas ethyl, propyl, allyl, hydroxyl groups, etc., are employed as sensitizers for black-and-white materials, acid nitrogen substituents are the principal substances which are built into the dyes which are used for sensitizing colour film, since acid substituents such as alkyl carboxylic acids or alkyl sulphonic acids render the dyes very readily adsorbable to the silver halide grains, and decrease the displacement of the sensitizers from the surface of the grains, an effect which is often produced notably in colour film by the colour couplers. At the same time the removal of dyes from the emulsions by washing is facilitated, and the risk of any coloration remaining behind in the film or paper is thus avoided.

A fact to be noted is that the introduction of *one* acid substituent into a cationic dye results in a Betain cyanine [35], and the introduction of *two* acid substituents gives rise to an anion [36]; cf. 150.

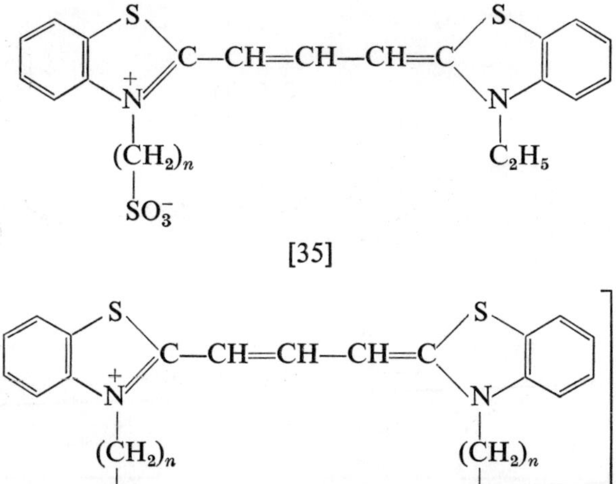

[35]

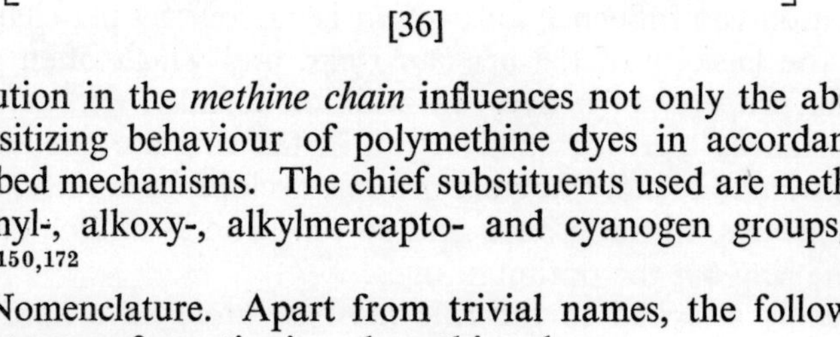

[36]

3. Substitution in the *methine chain* influences not only the absorption but also the sensitizing behaviour of polymethine dyes in accordance with the above-described mechanisms. The chief substituents used are methyl and ethyl groups, phenyl-, alkoxy-, alkylmercapto- and cyanogen groups, as well as halogens.[130,150,172]

2.3.1.1.2 Nomenclature. Apart from trivial names, the following nomenclature is customary for cationic polymethine dyes:

1. The methine cyanines are designated by the prefixes mono, tri, penta, etc., according to the number of methine groups situated between the two heterocyclic nuclei. Although, strictly speaking, one of the carbon atoms in each of the heterocyclic rings also count as carbon atoms *in the* polymethine chain and thus help to determine the position of maximum absorption, they are not taken into consideration when assigning the name to the compound.

For instance, the first dye in this class to be produced, namely the cyanine [37] which was synthesized by *Williams* in 1856, represents a monomethine dye. The absorption band is, however, situated comparatively far in the long wavelength region (at 597 mμ) since a heptamethine chain is present.

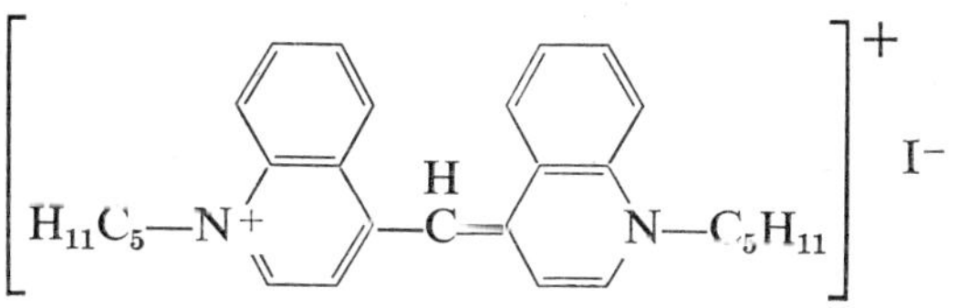

[37]

2. Monomethine dyes are often described as cyanines, trimethine dyes as carbocyanines, pentamethine dyes as dicarbocyanines, heptamethine dyes as tricarbocyanines, etc. For instance, the infra-red sensitizer with a maximum sensitization at 10,500 Å and the formula shown in [38] can be described as 3,3'-diethylthiapentacarbocyanine perchlorate or as undecacarbocyanine with 11 CH— groups between the benzothiazole nuclei.[88,173]

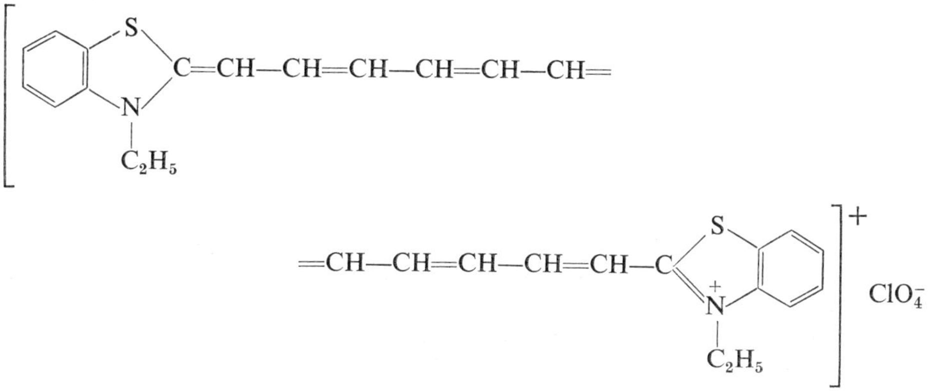

[38]

3. An exact formulation usually contains data of the ring systems at the ends of the methine chain as well as of the size of the intermediate chain. For example, the well-known Kryptocyanine (or Rubrocyanine) [39] the maximum

sensitization of which is at 735 mμ, is precisely *bis*-(1-ethylquinolyl-4)-trimethine cyanine iodide.

$$\left[H_5C_2-N^+ \diagup\!\!\!\diagdown -CH=CH-CH= \diagup\!\!\!\diagdown N-C_2H_5 \right]^+ \quad I^-$$

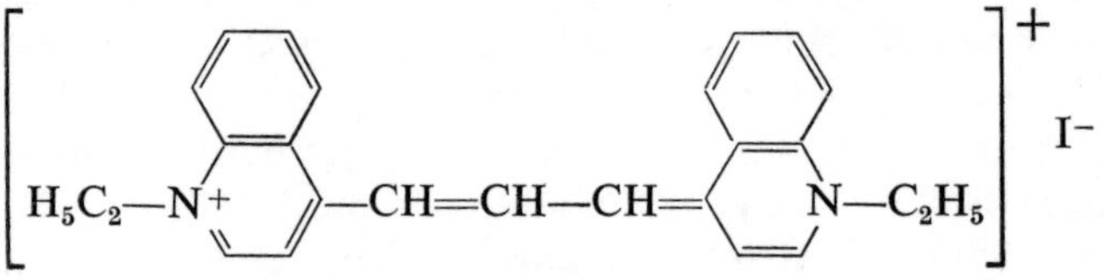

[39]

Analogously Pinacyanol [30] is *bis*-(1-ethylquinolyl-2)-trimethine cyanine iodide. The nomenclature used to describe the nuclei is often abbreviated; thus thia stands for benzothiazole, oxa for benzoxazole, selena for benzoselenazole, etc.

4. When the ring system at the ends of the chain is quinoline the following data are also to be found, depending on the point of attachment of the methine chain:

Pseudocyanine: for 2,2′-monomethine cyanine, e.g. Pinacyanol [30]
Cyanine: for 4,4′-monomethine cyanine, e.g. Kryptocyanine [39]
Isocyanine: for 2,4′-monomethine cyanine, e.g. Ethyl red [40]

which instead of (1-ethylquinoline-2)-(1′-ethylquinoline-4′)-monomethinecyanine iodide is briefly described as 1,1′-diethylisocyanine iodide.

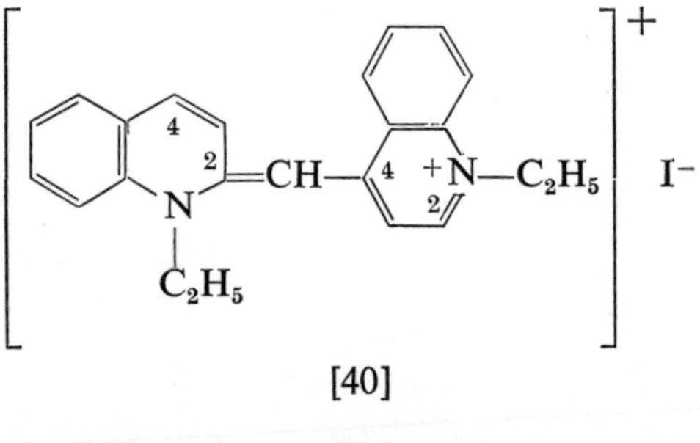

[40]

2.3.1.1.3 Synthesis. It is not intended to discuss here the various methods of preparing polymethine dyes, as this is beyond the scope of the present monograph. For this we would refer the reader to the work of *Riester, Dieterle, Brooker, Scheibe et al.*, which has already been cited, and to the reviews on the subject.[64,169]

We shall however sketch briefly two methods of synthesis which can be used:

Firstly, trimethine cyanines can easily be synthesized by the methods described by *König et al.*,[174,175] namely by condensing activated quaternary salts of heterocyclic nuclei with ortho esters such as *o*-formic ester $HC(OC_2H_5)_3$ in accordance with the following scheme.

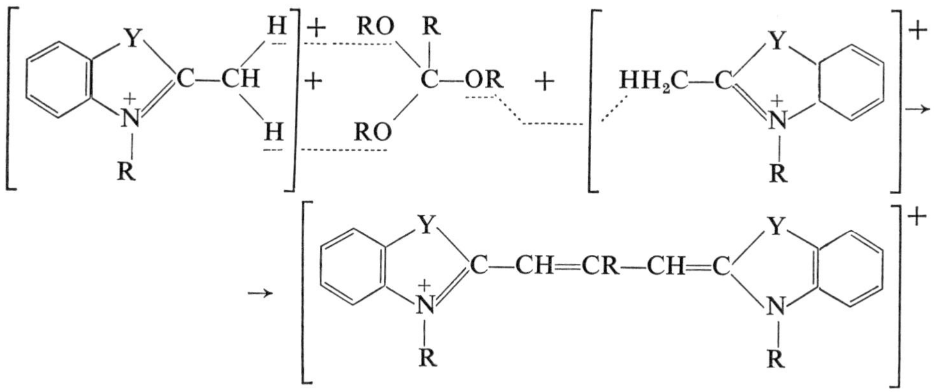

$$\left[\begin{array}{c}\includegraphics\end{array}\right]^+ + \begin{array}{c}RO\quad R\\ C-OR\\ RO\end{array} + \left[\begin{array}{c}HH_2C-C\end{array}\right]^+ \rightarrow$$

$$\rightarrow \left[\begin{array}{c}C-CH=CR-CH=C\end{array}\right]^+$$

Secondly, pentamethine cyanines and higher vinylene homologues are prepared by reacting quaternary salts of the heterocyclic nuclei with strepto-penta- and streptohepta-cyanines [33], etc., as illustrated in the following scheme; cf. Dieterle et al.[85,87,178]

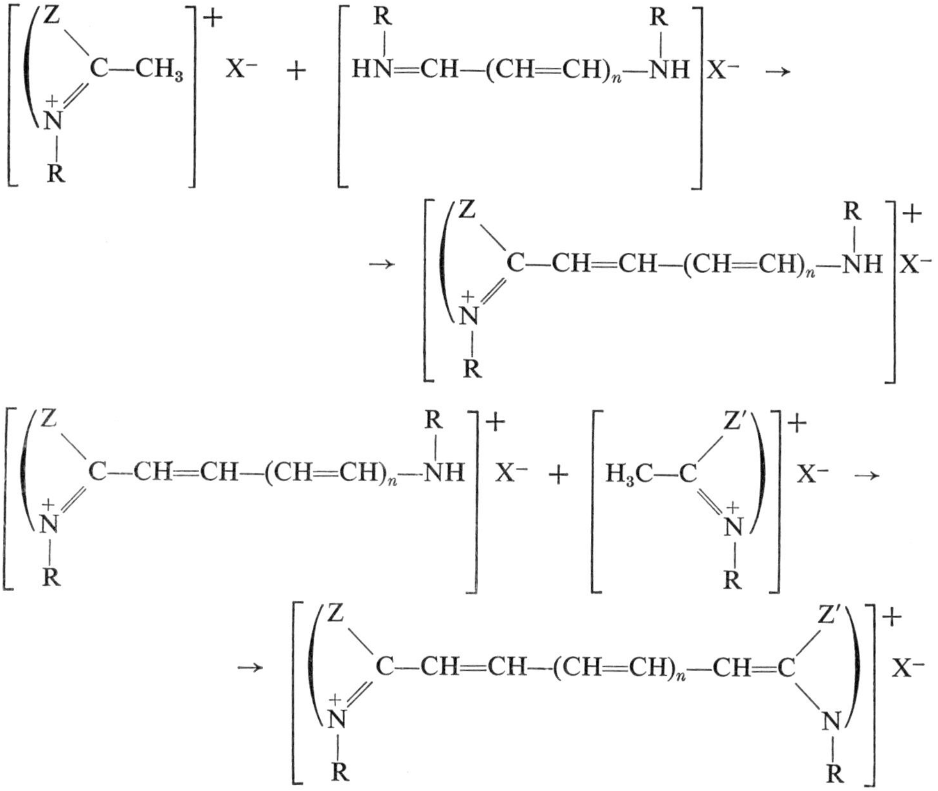

$$\left[\begin{array}{c}Z\\ C-CH_3\\ \overset{+}{N}\\ R\end{array}\right] X^- + \left[\begin{array}{c}R\qquad\qquad R\\ HN=CH-(CH=CH)_n-NH\end{array}\right] X^- \rightarrow$$

$$\rightarrow \left[\begin{array}{c}Z\qquad\qquad\qquad\qquad\qquad R\\ C-CH=CH-(CH=CH)_n-NH\\ \overset{+}{N}\\ R\end{array}\right]^+ X^-$$

$$\left[\begin{array}{c}Z\qquad\qquad\qquad\qquad R\\ C-CH=CH-(CH=CH)_n-NH\\ \overset{+}{N}\\ R\end{array}\right]^+ X^- + \left[\begin{array}{c}Z'\\ H_3C-C\\ \overset{+}{N}\\ R\end{array}\right]^+ X^- \rightarrow$$

$$\rightarrow \left[\begin{array}{c}Z\qquad\qquad\qquad\qquad\qquad\qquad Z'\\ C-CH=CH-(CH=CH)_n-CH=C\\ \overset{+}{N}\qquad\qquad\qquad\qquad\qquad\qquad N\\ R\qquad\qquad\qquad\qquad\qquad\qquad\qquad R\end{array}\right]^+ X^-$$

It should also be noted that the phosphorus analogues of the cyanines, namely the phosphinines, are also used as sensitizers.[179,180]

2.3.1.2 *Anionic or Acid Polymethine Dyes.* The chromophore chain of anionic dyes usually ends in oxygen atoms, so that the following mesomeric general formulae are characteristic of the dye salts.

$$\left[\overset{\ominus}{O}-\overset{|}{C}=\left(\overset{|}{C}-\overset{|}{C}=\right)_n O\right]^- X^+ \rightleftharpoons \left[O=\overset{|}{C}-\left(\overset{|}{C}=\overset{|}{C}-\right)_n \overset{\ominus}{O}\right]^- X^+$$

These include dyes which are often termed oxonoles, namely Fluorescein and its derivatives (Erythrosine, etc.) as well as the symmetrical and unsymmetrical polymethine dyes obtained by the combination of suitable heterocyclic nuclei —see Table 6.

TABLE 6

RING SYSTEMS OF THE OXONOLES

System	Structure	System	Structure
Rhodanine	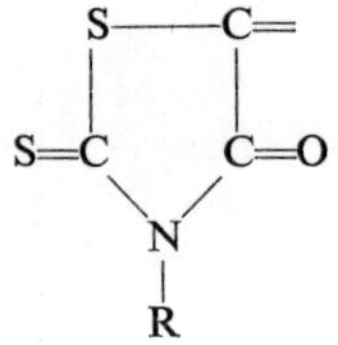 (S—C=, S=C, C=O, N–R ring)	Thiooxazolidine-dienone	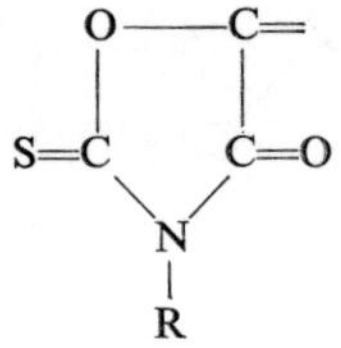 (O—C=, S=C, C=O, N–R ring)
Thiohydantoin	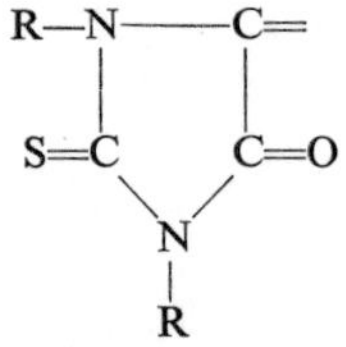 (R—N—C=, S=C, C=O, N–R ring)	Pyrazolin-5-one	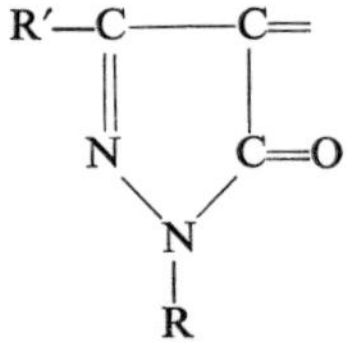 (R'—C—C=, N, C=O, N–R ring)

An example of this is given by the oxonol [41] which is synthesized from N-ethylrhodanine, and which corresponds to a trimethine cyanine

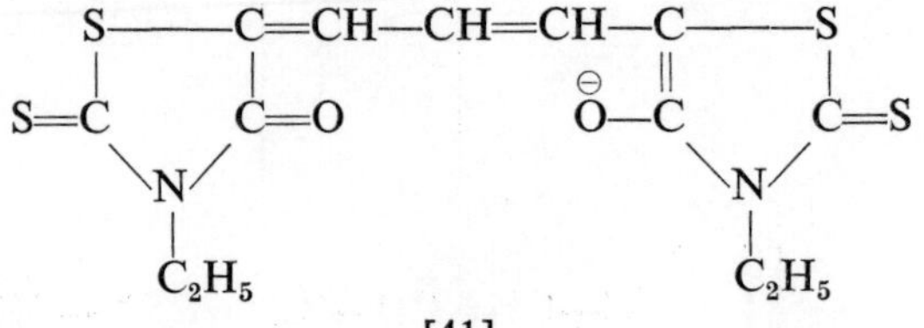

[41]

Acid polymethine dyes are also obtained by combining a nucleus (rhodanine, etc.) which participates in the structure of the oxonole, with a nitrile group (or two nitrile groups) via a chain containing an odd number of conjugated carbon atoms. One of these nitrile dyes[150] is, for example, dye [42]

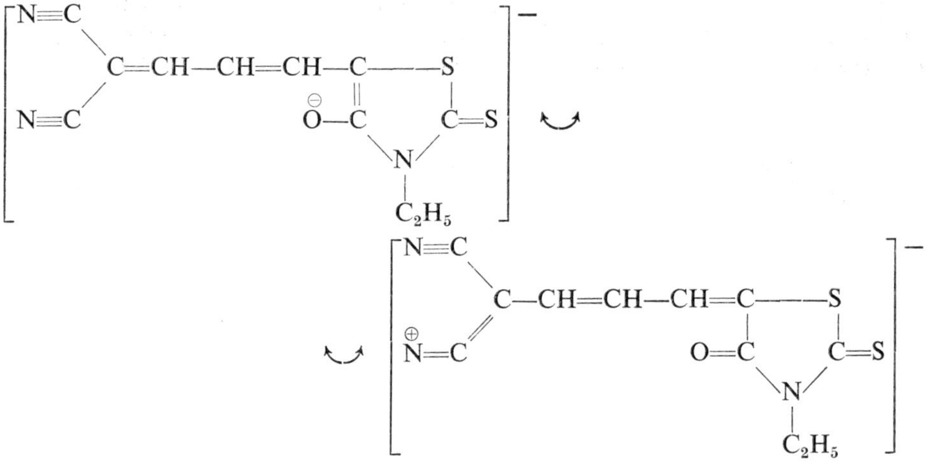

[42]

2.3.1.3 *Neutral Polymethine Dyes.* Nonionic (neutral) polymethine dyes are obtained by making the conjugated chain terminate in nitrogen and oxygen atoms or in heterocyclic ring systems containing positive and negative charges. Whereas one of the mesomeric general forms is uncharged the other is strongly dipolar in character, so that they can be described as intramolecular ionic dyes.

The neutrocyanines which, particularly in the case of the cyclic derivatives, are termed merocyanines, include many very good sensitizers. The heterocyclic ring systems of cationic and anionic polymethine dyes, linked together by a methine chain, enter into the structure of merocyanines (see Tables 5 and 6).

As an example of this group which only contains asymmetrical dyes we shall quote here a dimethineneutrocyanine [43] which consists of a benzothiazole nucleus and a rhodanine nucleus:

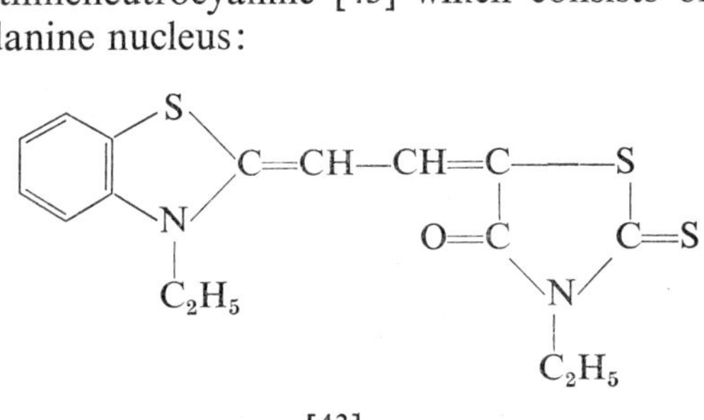

[43]

For methods of synthesizing these sensitizers cf. *Riester et al.*[169,181–183] In the case of the specified benzothiazolerhodaninedimethine-neutrocyanine [43] it is possible first to convert rhodanine for example into an intermediate product by means of an orthoester, and then to condense this intermediate product with the quaternary salt of benzothiazole to form a merocyanine.[184]

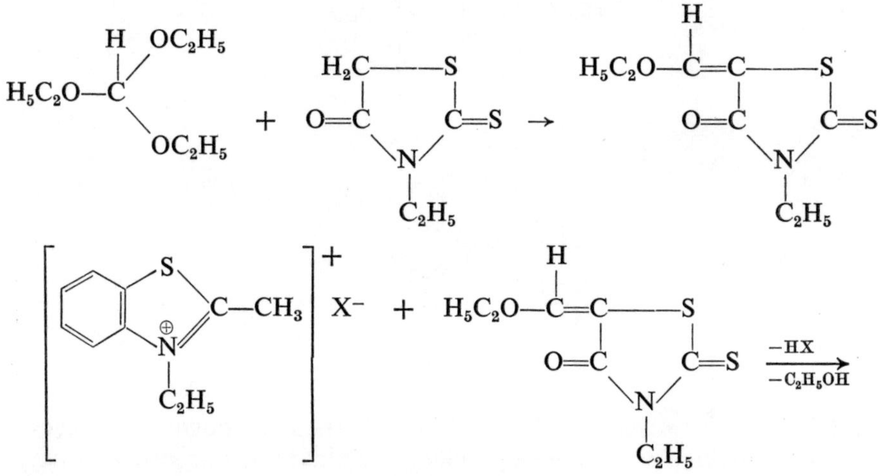

→ Merocyanine [43]

Similar methods can also be used for synthesizing other merocyanines. Thus, tetra- and hexa-methine-neutrocyanines can be formed by using the quaternary heterocyclic salts obtained by reaction with streptopolymethine cyanines.

In addition to the merocyanines, the neutrocyanines also include the hemi-oxonoles, in which the basic component is not enclosed in a ring. Moreover, by combining a basic nucleus with a nitrile group, neutral nitrile dyes are obtained, and if the compounds are synthesized using phosphorus compounds

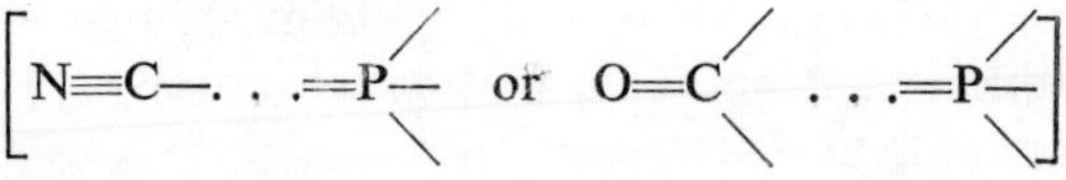

phosphinines are obtained.[185]

2.3.2 POLYMETHINE DYES OF HIGHER ORDER. The polymethine dyes of higher order include a series of dyes of complex structure, for which the following classification, as given by *Zwicky*,[150] is chosen.

2.3.2.1 *Rhodacyanines*. Rhodacyanines are basic cyanines of higher order in which a merocyanine is combined with a cyanine. As the dye [44] which is given as an example shows, these dyes contain a rhodanine ring

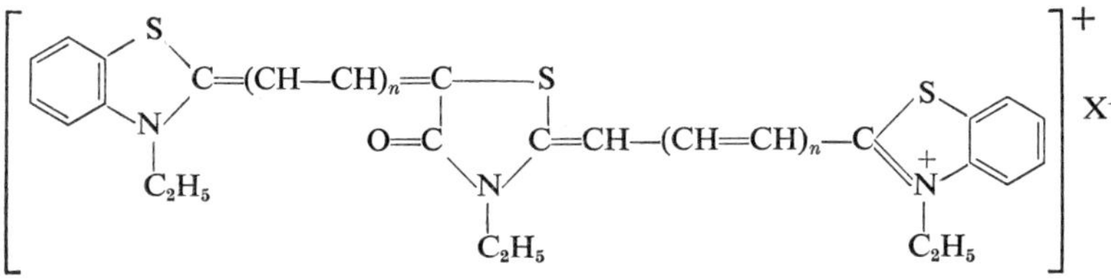

[44]

built into the methine chain. Other ring systems such as thiohydantoin or hydantoin can, however, also be built into the chain. The ring system can also be extended to give dyes containing four or more nuclei.

The incorporation of rings gives rise on the one hand to bathochromic displacements which, in the case of rhodanine, are situated at 50 mμ. On the other hand, a distinctive feature of these dyes is their very narrow region of sensitization, which, by varying the number of methine groups, can be established in any part of the spectrum to suit the type of sensitization required. In addition, these dyes are strongly resistant to the action of couplers, so that they find an important use in colour emulsions.[186]

2.3.2.2 *The Neocyanine Type of Dye.* The neocyanine type of dye [45] includes trinuclear dyes, which according to the neocyanine framework, are built up of similar or different nuclei.[188-190, 190a]

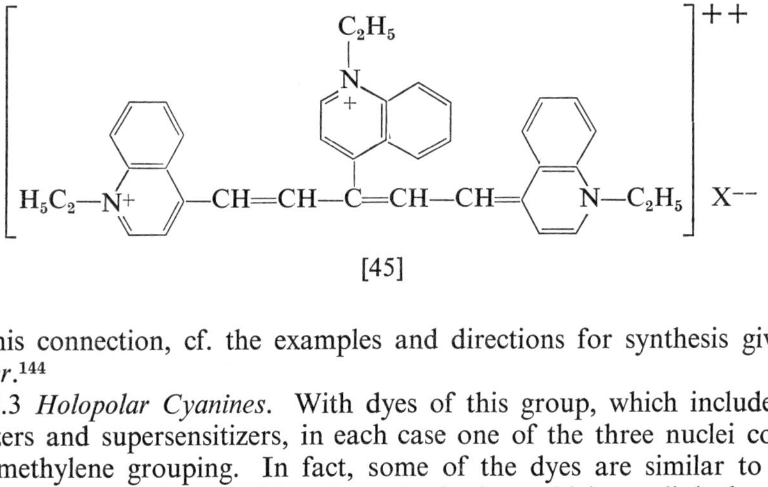

[45]

In this connection, cf. the examples and directions for synthesis given by *Brooker*.[144]

2.3.2.3 *Holopolar Cyanines.* With dyes of this group, which include good sensitizers and supersensitizers, in each case one of the three nuclei contains a ketomethylene grouping. In fact, some of the dyes are similar to merocyanines, and contain an acid and two basic rings which are linked together by means of methine groups. Since, for spatial reasons, the debatable general structures I–III of the given dye [46] cannot all be involved simultaneously in the mesomeric ground state, this type of dye yields two stereoisomeric modifications, which among other things, differ in their absorption behaviour.[161]

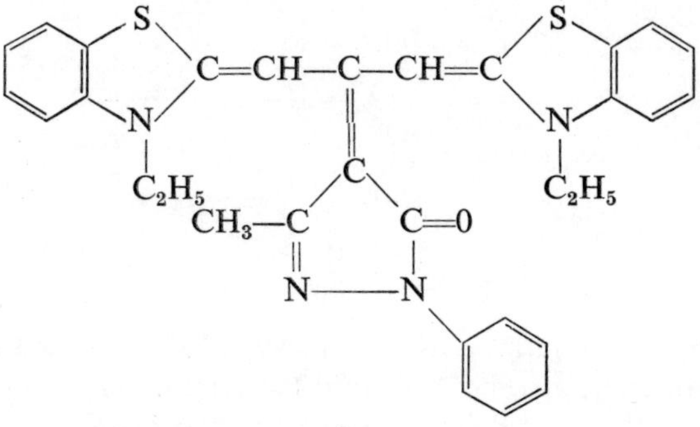

I

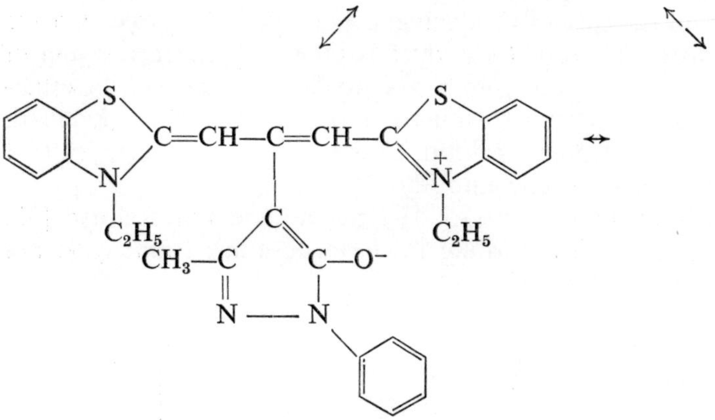

II

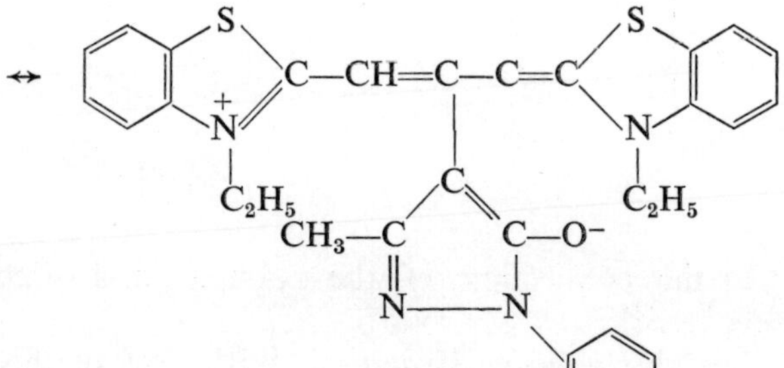

III

[46]

In this connection it should also be pointed out that mesomerism is generally taken to signify the effect of the overlapping of various, similar electron structures in a system, with the formation of a new, stable electron configuration.

In order to express this mesomeric state (the ground state of a compound) which is brought about by the interaction of the π-electrons, as a formula, the electron structures which enter into reciprocal action with each other (general structures) are given.[53] General structures which are particularly rich in energy do not however participate in the mesomeric state.

Holopolar form: Mesomerism between nuclei which are situated in one plane can only be achieved by rotating the acid ring system out of the plane of the two basic nuclei, i.e. II⇌III. The positive charge on the cationic polymethine is thereby neutralized by the negative charge on the anionic group which lies in a different plane.

In the example given, the dye isomer therefore corresponds to a cationic benzothiacarbocyanine with a pyrazolonate anion as the substituent in the chain. This representation is in harmony with the fact that the position of the absorption band corresponds with that of benzothiacarbocyanine. It must, however, be borne in mind that the isomeric state which is due to the complete separation of the charges—hence the name holopolar—is only slightly stable. The holopolar form is therefore only obtained in strongly polar solvents (water, and lower alcohols such as methanol).[191]

Meropolar form: The acid anionic nucleus in this isomeric modification occurs in the same plane as one of the basic nuclei, and therefore in a state of mesomerism, whereas the second basic ring system is rotated out of the plane. The dye is therefore a merocyanine with an uncharged (I) and a charged (II or III) mesomeric general structure. In example [46]—in agreement with the absorption—this corresponds to a benzothiazolepyrazolone-neutrocarbocyanine with a benzothiazole group substituted in the chain. As the meropolar isomeric modification exhibits only slight polarization and is therefore not so unstable as the holopolar form, the "meropolar" absorption band at the shorter wavelengths is obtained even in non-stabilizing solvents such as methylcyclohexane.

The stereoisomeric effect which is described here and which is indicated by a perceptible displacement in absorption on changing over to strongly polar solvents, is called *allopolar isomerism*. In this connection it may be noted that cis-trans isomers likewise behave differently in regard to their sensitizing power.[149,192]

2.3.3 THE PREPARATION OF SENSITIZED EMULSIONS. There are two methods of sensitizing photographic emulsions. Firstly, the dye which is dissolved in methanol or in any other solvent which is miscible with water (about 1 : 1000) can be stirred into the emulsion at 30°–40°C at the end of the manufacturing process before coating.[169] Secondly, the finished photographic emulsion layers can be sensitized by bathing them in a very dilute solution of dye (about 1 : 10^6).[193,194] Since the layers of sensitizer obtained by bathing are frequently unevenly adsorbed and moreover the emulsions often exhibit poor keeping properties, this method is seldom used nowadays.

Various factors have to be taken into consideration when determining the optimum quantities of sensitizer to be added. With very small quantities of dye the photographic sensitivity in the region of sensitization first increases in proportion to the amount of dye and finally approaches a limit. On reaching this limit, which corresponds to a surface coverage of about 60 to 80%, the sensitivity remains largely constant in the region of optimum sensitization even when the quantity of dye is changed. Therefore in practice it is advisable to use a quantity of dye which corresponds to this boundary, since in the region of proportionality small variations in concentration result in too great a change in the sensitizing action.

The use of sensitizers in too large quantities, thus has to be avoided, as in certain circumstances this can give rise not only to a reduction in the optimum sensitizing action, but it can also release a more or less well-defined desensitizing effect.

Another important point to be taken into consideration in practice is that the optimum quantity of sensitizer is dependent on the constitution of the dye, the optimum quantity decreasing with increase in the length of the methine chain. Whereas, say, 200 mg of dye per mol of AgBr are needed for optimum sensitization when carbocyanines are used, in the case of pentamethine cyanines this quantity amounts to about 60 mg/mol of AgBr and with nonamethine-cyanines to only 6 mg/mol of AgBr.[64] Since, moreover, the tendency observed for dyes to desensitize with overdosage increases with the length of the chain, the region of optimum sensitization is smaller in the case of sensitizers which absorb at long wavelengths than for those which absorb at short wavelengths. For this reason with infra-red sensitizers, the specified quantities which are of the order of only a few mg must be strictly adhered to.

It should also be noted that the spectral sensitization of electrophotographic layers, e.g. on a ZnO support, is achieved by mixing ZnO with dye solutions, toluene and a silicone resin[50] or by adding the solution to the ZnO layer.[195] Since cyanine dyes are adsorbed less well to ZnO than to silver halides, concentrated solutions have to be used. For information on the problem of removing dyes from an exposed layer, cf. *Eastman*.[196]

An analogous procedure is followed for sensitizing the layers intercalated in photoelectric cells. In the case of ZnO transverse field cells, the ZnO powder is, for instance, stirred into a paste with a dye (Pinaflavol, Rose Bengals, etc.) dissolved in ethyl alcohol, and dried in a thin layer of about 20 μ in thickness on the base of the cell.[197]

References

[80] VOGEL, W. H., *Ber. dtsch. chem. Ges.*, **6**, 1302 (1873).

[81] BECQUEREL, E., *C.R. hebd. Séances Acad. Sci.*, **79**, 185 (1874).

[82] PLOTNIKOW, J., *Allgemeine Photochemie (General Photochemistry)*, W. de Gruyter (Berlin-Leipzig) 1936.

[83] U.S.P. 844,804 (1907); **172**, 118 (1906) B. Homolka.

[84] ADAMS, E. Q. and HALLER, H. L., *J. Amer. chem. Soc.*, **42**, 2661 (1920).
[85] SCHEIBE, G., *Dissertation (Thesis)*, Universität Erlangen 1918.
[86] U.S.P. 1,804,674, H. T. CLARKE (1931).
[87] BROOKER, L. G. S., HAMER, F. M. and MEES, C. E. K., *J. opt. Soc. Amer.*, **23**, 216 (1933).
[88] DIETERLE, W. and RIESTER, O., *Z. wiss Photogr.*, **36**, 68 and 141 (1937).
[89] U.S.P. 2,756,227 and U.S.P. 2,734,900, D. W. HESELTINE to Eastman Kodak Co. (1953).
[89a] LEERMAKERS, J. A., *J. chem. Phys.*, **5**, 878 (1937).
[90] ZWICKY, H., *Chimia*, **15**, 300 (1961).
[91] WEST, W. and GEDDES, A. L., *J. Phys. Chem.*, **68**, 837 (1964).
[92] LEVKOJEFF, I. I., LIFSCHITZ, E. B., NATANSON, S. W., SWESCHNIKOW, N. N. and SITNIK, Z. P., *Wissenschaftliche Photographie (Scientific Photography)*, (Int. Conf. Köln, 1956), edited by O. HELWICH, p. 109, (Darmstadt) 1958.
[93] SCHEIBE, G., *Angew. Chem.*, **50**, 51 and 212 (1937); **52**, 631 (1939).
[94] JELLEY, E. E., *Nature, Lond.*, **138**, 1009 (1936).
[95] SCHEIBE, G., *Z. Elektrochem.*, **52**, 283 (1948); *Kolloidzschr.*, **82**, 1 (1938).
[96] SCHEIBE, G., *Z. Elektrochem.*, **47**, 73 (1941).
[97] FÖRSTER, TH. and KÖNIG, E., *Z. Elektrochem.*, **61**, 344 (1957).
[98] CARROLL, B. H., *Scientific Photography*, Proc. Int. Colloq. Liége 1959, edited by H. SAUVENIER, p. 427, Pergamon Press (Oxford) 1962.
[99] RABINOVITCH, E. and EPSTEIN, L., *J. Amer. chem. Soc.*, **63**, 69 (1941).
[100] VICKERSTAFF, T. and LEMIN, D. R., *Nature, Lond.*, **157**, 373 (1946).
[101] ZANKER, V., *Z. phys. Chem.*, **199**, 225 (1952); *Z. phys. Chem. (N.S.)*, **2**, 52 (1954); **8**, 20 (1956).
[101a] EMERSON, E. S., CONLIN, M. A., ROSENOFF, A. E., NORLAND, K. S., RODRIGUEZ, H., CHIN, D., and BIRD. G. R., *J. phys. chem.* **71**, 2396 (1967).
[101b] MCKAY, R. B., and HILLSON, P. J., *Trans. Faraday Soc.*, **61**, 1800 (1965).
[102] SCHEIBE, G., MAREIS, A. and ECKER, G., *Naturwissenschaften*, **25**, 474 (1937).
[103] HOPPE, W., *Kolloidzschr.*, **101**, 300 (1942).
[104] JELLEY, E. E., *Nature, Lond.*, **139**, 631 (1937).
[105] KÄUFFER, H. and SCHEIBE, G., *Z. Elektrochem.*, **59**, 584 (1955).
[106] SCHEIBE, G., MÜLLER, R. and SCHIFFMANN, R., *Z. phys. Chem.*, **B49**, 324 (1941).

[107] SKERLAK, T., *Kolloidzschr.*, **95**, 265 (1941).
[108] ZIMMERMANN, H. and SCHEIBE, G., *Z. Elektrochem.*, **60**, 566 (1956).
[109] FRIESER, H., GRAF, A. and ESCHRICH, D., *Z. Elektrochem.*, **65**, 870 (1961).
[110] BORGINON, H. and DANKAERT, V., *Photogr. Korr.*, **98**, 74 (1962).
[111] BOYER, S. and CAPPELAERE, J., *J. Chim. phys.*, **60**, 1123 (1963).
[112a] OLLMANN, D. and PIETSCH, H., *Veröff. Wiss. Photo-Lab. Wolfen*, **10**, 87 (1965).
[112] WOOD, H. W., *Sci. Ind. Photogr.*, (2), **25**, 465 (1954).
[113] DICKINSON, H. O., *Photogr. J.*, **90B**, 142 (1950).
[114] MEYER, K. and KUNZE, K., *Z. wiss. Photogr.*, **53**, 209 (1959).
[115] MATTOON, R. W., *J. chem. Phys.*, **12**, 268 (1944).
[116] SHEPPARD, S. E., *Science*, **93**, 42 (1941).
[117] U.S.P. 2,373,659, B. H. CARROLL and J. SPENCE to Eastman Kodak Co. (1945).
[118] LEROY, G. and NYS, J., *Bull. Soc. Chim. Belges.*, **73**, 673 (1964).
[119] SCHMIDT, O., *Z. Elektrochem.*, **43**, 238 (1937); *Chem. Ber.*, **73**, 97 (1940).
[120] SOMMERFELD, A., *Z. Phys.*, **47**, 1 (1928).
[121] KUHN, H., *J. chem. Phys.*, **16**, 287 (1948); **17**, 1198 (1949); **34**, 1308 (1951).
[122] KUHN, H., *Angew. Chem.*, **71**, 93 (1959).
[123] KUHN, H., *Helv. chim. acta*, **31**, 1441 (1948).
[123a] DÄHNE, S., *Z. Chem.*, **5**, 444 (1965).
[123b] KUHN, H., *Helv. chim. acta*, **32**, 2247 (1949); *Chimia (Zürich)*, **4**, 203 (1950).
[123c] HÜNIG, S. and QUAST, H., *Optische Anregung organischer Systeme (Optical Excitation of Organic Systems)*, edited by W. FOERST, p. 184, Verlag Chemie (Weinheim/Bergstr.) 1966.
[123d] FRIEND, J. A. and ROBERTS, N. K., *Austr. J. Chem.*, **11**, 104 (1958).
[123e] LEUPOLD, D. and DÄHNE, S., *Z. phys. Chem.*, **48**, 24 (1966).
[123f] DÄHNE, S., *Mber. dtsch. Akad. Wiss*, **5**, 454 (1963).
[123g] DÄHNE, S. and ALTHOFF, H., *Z. wiss Photogr.*, **58**, 69 (1964).
[124] BROOKER, L. G. S., *Rev. mod. Phys.*, **14**, 275 (1942).
[125] BROOKER, L. G. S., KEYES, G. H. and WILLIAMS, W. W., *J. Amer. chem. Soc.*, **64**, 199 (1942).
[126] HAMER, F. M., *The Cyanine Dyes and related Compounds*, p. 688, Interscience Pub. (New York, London) 1964.

[127] HAMER, F. M., *J. chem. Soc.*, **125**, 1348 (1924).
[128] FISHER, N. T. and HAMER, F. M., *J. chem. Soc.*, 907 (1937).
[129] KNOTT, E. B., *J. chem. Soc.*, 1024 (1951).
[130] RIESTER, O., *Mitt ForschungsLab Agfa Leverkusen-München*, Vol. I, p. 44, Springer (Berlin) 1955.
[131] KIPRIANOV, A. I., *Ukr. Khim. Zh.*, **20**, 204 (1954).
[132] KUHN, H., *Helv. chim. acta*, **34**, 2371 (1951).
[133] HÜCKEL, H., *Z. Elektrochem.*, **43**, 752 (1937).
[134] ZWICKY, H., *Chimia*, **9**, 37 (1955).
[135] MEES, C. E. K., *Theory of the Photographic Process*, Macmillan (New York) 1955.
[136] PIANKA, M., BARANY, H. C. and SMITH, C. G., *Nature, Lond.*, **167**, 440 (1951).
[137] BARANY, H. C. and PIANKA, M., *J. chem. Soc.*, 2217 (1953).
[138] LEVKOEV, I. I., SVESHNIKOV, N. N. and KHEIFETS, S. A., *Zh. obshch. Khim.*, **16**, 1489 (1946).
[139] LEVKOEV, I. I. and PORTNAYA, B. S., *Zh. obshch. Khim.*, **21**, 2050; Engl. transl. 2287 (1951).
[140] FISHER, N. I. and HAMER, F. M., *Proc. roy. Soc.*, Ser. A, **154**, 703 (1936).
[141] HAMER, F. M. and KELLY, M. I., *J. chem. Soc.*, 777 (1931).
[142] STAAB, H. A., *Einführumg in die theoretisiche organische Chemie*, (*Introduction to Theoretical Organic Chemistry*), Verlag Chemie (Weinheim) 1959.
[143] BROOKER, L. G. S., WHITE, F. L. and SPRAGUE, R. H., *J. Amer. chem. Soc.*, **73**, 1087 (1951).
[144] BROOKER, L. G. S., *The Theory of the Photographic Process*, 3rd edn., edited by C. E. K. MEES and T. H. JAMES, p. 198, Macmillan (New York, London) 1966.
[145] BROOKER, L. G. S., SKLAR, A. L., CRESSMAN, H. W. J., KEYES, G. H., SMITH, L. A., SPRAGUE, R. H., VAN LARE, E., VAN ZANDT, G., WHITE, F. L. and WILLIAMS, W. W., *J. Amer. chem. Soc.*, **67**, 1875 (1945).
[145a] SYKES, A., *Reaktionsmechanismen der organischen Chemie*, (*Mechanisms of the Reactions in Organic Chemistry*), p. 46 *et seq.*, Verlag Chemie (Weinheim/Bergstr.) 1966.
[146] LOWER, S. K. and EL-SAYED, M. A., *Chem. Rev.*, **66**, 199 (1966).
[147] CALVERT, J. G. and PITTS, J. N., *Photochemistry*, p. 244 *et seq.*, John Wiley & Sons (New York) 1966.

[148] WEST, W., *J. phys. Chem.*, **66**, 2398 (1962).
[149] ANDERSON, G. DE W., *Scientific Photography*, Proc. Int. Colloq. Liége 1959, edited by H. SAUVENIER, p. 487, Pergamon Press (Oxford) 1962.
[150] ZWICKY, H., *Chimia*, **19**, 416 (1965).
[151] U.S.P. 2,503,776, R. H. SPRAGUE, to Eastman Kodak Co. (1950).
[152] Brit. P. 904,322, J. M. NYS and H. DE POORTER, to Gevaert Photo-Producten, N.V. (1962).
[153] G.F.P. 929,080, O. RIESTER, to Agfa AG, (1955).
[154] U.S.P. 2,256,163, K. KUMETAT and G. WILMANNS, to General Aniline and Film Corp., (1941).
[155] U.S.P. 2,282,116, L. G. S. BROOKER, to Eastman Kodak Co. (1942).
[156] KENDALL, J. D., *IXe Congrès Intern. de Photographie Scientifique et Appliquée, Paris, 1935*, (*IXth Internat. Congress of Scientific and Applied Photography, Paris, 1935*), Editions Rev. d' Optique, p. 227 (Paris) 1936.
[157] FUCHS, K. and GRANANG, E., *Ber. dtsch. chem. Ges.*, **61**, 57 (1928).
[158] SCHEIBE, G. and DÖRR, F., *Scientific Photography*, Proc. Internat. Colloq. Liége 1959, edited by H. SAUVENIER, p. 140, Pergamon Press (New York) 1962.
[159] SHEPPARD, S. E., LAMBERT, R. H. and WALKER, R. D., *J. chem. Phys.*, **9**, 107 (1941).
[160] BROOKER, L. G. S., *Experientia* (Basel), Supplement II/**16**, 229 (1955).
[161] BROOKER, L. G. S., WHITE, F. L., HESELTINE, D. W., KEYES, G. H., DEUT, S. G., JR. and VAN LARE, E. J., *J. photogr. Sci.*, **1**, 173 (1953).
[162] BRUYLANTS, P., VAN DORMAEL, A. and NYS, J. M., *Bull. Cl. Sci. Acad. roy. Belgique* (5), **34**, 703 (1948).
[162a] GRECHKO, M. K., NATANSON, S. V. and ALPEROVICH, M. A., *Kinotekh., Nauch.-Tekh.*, **4**, 92 (1963).
[163] NATANSON, S. V. and KLIMSO, E. F., *Zh. nauchn. prikl. Foto. Kinematogr.*, **5**, 452 (1960).
[163a] FRIESER, H., GRAF, A. and ESCHRICH, D., *Z. Elektrochem.*, **64**, 1107 (1960).
[164] MARKOCKI, W., *J. photogr. Sci.*, **13**, 85 (1965).
[165] MARKOCKI, W. and ROMER, W., *Korpuskularphotographie*, IV, 194 (1963).
[166] BOKINIK, YA. I. and ILJINA, A. Z., *Acta Phys. Chem.*, **9**, 205 (1928).
[167] PADDAY, J. F. and WICKHAM, R. S., *Trans. Faraday Soc.*, **62**, 1283 (1966).

[168] GÜNTHER, E. and MOISAR, E., *J. photogr. Sci.*, **13**, 280 (1965).
[168a] EGGERS, J., GÜNTHER, E. and MOISAR, E., *Photogr. Korr.*, **102**, 144 (1966).
[169] RIESTER, O., *Ullmanns Encyklopädie der technischen Chemie*, (*Ullmanns Encyclopedia of Industrial Chemistry*), 3rd edn., edited by W. FOERST, Vol. 14, Urban & Schwarzenberg (Munich-Berlin) 1963.
[170] DUFFIN, G. F., *Rep. Progr. appl. Chem.*, **42**, 218 (1957).
[171] CIERNICK, J., MISTR, A. and VÁVROVÁ, J., *IV. wiss angew fotogr. Konf. Budapest*, (*IVth Conf. on sci. appl. Phot. Budapest*), 1963, 237, quoted from *Chem. Zbl.*, **137**, No. 35, 3370 (1966).
[172] FICKEN, G. E. and KENDALL, J. D., *Chimia*, **15**, 110 (1961)
[173] DIETERLE, W. and RIESTER, O., *Veröffentlichungen des wissenschaftlichen Zentral-Laboratoriums der phtographischen Abteilung-AGFA* (*Publications of the Photographic Department of the AGFA Central Scientific Laboratory*), Vol. V, p. 219, Hirzel (Leipzig) 1937.
[174] KÖNIG, W., *Ber. dtsch. chem. Ges.*, **55**, 3293 (1922); *J. pract. Chem.*, **109**, 324 (1925).
[175] HAMER, F. M., *J. chem. Soc.*, 3160 (1928).
[176] DIETERLE, W., DUER, H. and ZEH, W., *Veröffentlichungen des wissenschaftlichen Zentral-Laboratoriums der photographischen Abteilung-AGFA* (*Publications of the Photographic Department of the AGFA Central Scientific Laboratory*), Vol. III, p. 125, Hirzel (Leipzig) 1933.
[177] KÖNIG, W., *Z. wiss Photogr.*, **34**, 15 (1935).
[178] DIETERLE, W. and ZEH, W., *Z. wiss Photogr.*, **34**, 245 (1935).
[179] VAN DORMAEL, A., NYS, J. and DE POORTER, H., *Sci. Ind. Photogr.*, **31**, (2), 389 (1960); *Z. wiss Photogr.*, **54**, 152 (1960); *Photogr. Sci.*, **9**, 70 (1961).
[180] RAMIREZ, F. and LEVY, S., *J. Amer. chem. Soc.*, **79**, 6167 (1957).
[181] COENEN, M. and PESTEMER, M., *Z. Elektrochem.*, **57**, 785 (1953).
[182] KNOTT, E. B., *J. chem. Soc.*, 1482 (1954).
[183] D.A.S. 1,179,110, E.-J. POPPE and J. BRUNKEN (VEB Filmfabrik Agfa Wolfen) K. 57 b dated 22/7.1963, issued 1/10.1964.
[184] CHIEN-PE, LO and CROXALL, W. J., *J. Amer. chem. Soc.*, **76**, 4166 (1954).
[185] VAN DORMEAL, A., *Chimia*, **15**, 67 (1961).
[186] D.A.S. 1,177,481, O. RIESTER and M. GLASS (Agfa Akt. Ges., Leverkusen) K. 57 b dated 18/5.1963, issued 3/9.1964.
[187] G.F.P. 890,249, O. RIESTER, Farbwerke Hoechst (1937).
[188] HAMER, F. M., RATHBONE, R. J. and WINTON, B. S., *J. chem. Soc.*, 1434 (1947).
[189] KENDALL, J. D. and MAJER, J. R., *J. chem. Soc.*, 690 (1948).
[190] HAMER, F. M., RATHBONE, R. J. and WINTON, B. S., *J. chem. Soc.*, 1872 (1948).
[190a] KENDALL. J. D., and FRY, D. J., *Brit. P.* 544, 645, Oct. 16 (1940).
[191] BROOKER, L. G. S., *XIVth Intern. Congr. of Pure and Applied Chem.*, Zürich 1955, p. 229, Birkhäuser Verlag (Basel) 1955.
[192] ALPEROVICH, M. A., GRECHO, M. K. and NAUMOV, YU. A., *Zh. nauch. prikl. Foto Kinematogr.*, **8**, 410 (1963).
[193] DUNDON, M. L., *Amer. Photogr.*, **20**, 670 (1926).
[194] WALTHERS, F. M. and DAVIS, R., *Bur. Stand. Sci. Papers*, **422**, 353 (1921).
[195] NAMBA, S., *J. Phys. Chem.*, **69**, 774 (1965).
[196] Brit. P. 1,020,011 (Cl. G 03g), EASTMANN, D. R., Feb. 16 (1966).
[197] SHISHIDO, C., KIKUCHI, I., YONEZAWA, Y., WADA, M. and TAKAHASHI, T., *Sci. Rep. Ritu.*, B. **16**, 43 (1964).

III. THE FUNDAMENTALS OF SPECTRAL SENSITIZATION

3.1 General Laws Governing Photographic Sensitization

3.1.1 FACTORS CONCERNED WITH CONSTITUTION. The sensitizing power of a dye is dependent on various factors which are linked up with the constitution of the dye[134,150,198,199], and several of which have already been discussed in Chap. II. The essential conditions can be summarized in the following points.

1. In order to sensitize, the molecules must have a flat structure. The slightest deviations from coplanarity of the conjugation system are sufficient to decrease the sensitizing power of a dye, and here it is immaterial whether these deviations are due to the mutual overlapping of the substituents or to overlapping by the chain or a substituent in the chain, respectively[161]; cf. Chap. II, 2.4.

2. The dye must be firmly adsorbed to the surface of the silver halide grains. Without this close union between the dye and the silver halide grain it is impossible for the transfer of energy from the sensitizer to the AgBr, which forms the basis of the sensitization process, to be sufficiently exploited. Since this condition can easily be upset by the substances such as stabilizers, gelatin-hardening agents, wetting agents, etc. which are present in a photographic emulsion, a low intensity of sensitization can often also be attributed to a partial displacement of the dye from the surface of the grain.

3. The sensitizing action of dyes of similar constitution is greatly influenced by substituents[130]. For instance, the introduction of an atom (nitrogen, etc.) which acts as an electron-acceptor, or of a group into a position in a conjugated chain, which contains an odd number of members and in which positive charge has been induced by a key atom (auxochrome), destroys the sensitizing action of a dye. Moreover, this regression in the sensitizing power of a dye is often replaced by a desensitizing action; cf. Chap. II, 2.3.

4. The adsorbed dye must not enter into any reaction with the silver halide which can in any way adversely affect the course of the photographic process. For instance, fog formation, i.e. a reduction of silver ions in the dark or destruction by oxidation of the silver nuclei in the latent image formed on exposure, would come into this category.

Dyes with easily reducible functional groups (such as NO_2, etc.) are therefore not suitable as sensitizers. In this connection we would also refer to the problem presented by the destruction of sensitizing dyes by photolytic bromine[200].

3.1.2 PHYSICO-CHEMICAL LAWS. Photographic spectral sensitization conforms with a number of physico-chemical laws.

3.1.2.1 *Quantum Yield.* The absolute quantum yield ϕ of spectral sensitization is defined by the number of silver atoms formed photolytically for each quantum of light absorbed. In the case of good sensitizers it reaches a maximum of the order of magnitude of $\phi = 0.25$.[201,202]

Since the quantum yield of non-sensitized emulsions in the region of the normal absorption of silver halide is of the order of 1,[203-205] this signifies that part of the light-energy absorbed during spectral sensitization is not used for the photographic process. With less efficient sensitizers this portion might even be larger, as shown by the measurements made by *Wilcke et al.*[206,207] —ϕ up to 0·01. Moreover, in the region of the normal absorption of silver halide the sensitizers usually cause the quantum yield to fall from about $\phi = 1$ to $\phi = 0.8 - 0.7$.[207]

It must be remembered that because of the difficulty experienced in making absolute measurements of quantum yields, the efficiency of sensitization is generally stated in relation to the sensitivity in the unsensitized region[44]. In favourable cases this relative quantum yield ϕ_r^{450} can reach a value of 1;[98,201,208a] see Chap. I, 2.6.

3.1.2.2 *The repeated reactivity of the dye molecules.* During the course of sensitization a dye molecule is capable of forming at least 100 silver atoms,[166,203,208-211] before being destroyed by photochemical side reactions.

This finding has an important bearing on the theory of sensitization. For example it provides an argument against the formation of light-sensitive dye-silver halide compounds which are said to decompose into a silver atom and a dye residue on exposure,[212] since, in this case, the dye could only react once.

3.1.2.3 *Spectral sensitization in relation to temperature.* The photographic sensitivity of spectrally sensitized emulsions decreases to a greater extent with fall in temperature than it does in the case of non-sensitized emulsion layers.[109,213,214]

According to *Frieser et al.*,[109] this effect is characterized by the following regularities in behaviour.

1. From a certain temperature onwards there is a greater fall in the spectral sensitivity in the sensitized region than in the sensitivity of the silver halide in its region of absorption. Whereas the induced sensitivity falls by about one power of ten per 90°C., relatively little decrease occurs in the normal sensitivity up to $-180°$.

2. The desensitizing action of a dye likewise regresses at low temperatures, and for this reason, with strong desensitizers (Phenosafranine, etc.), the blue sensitivity can increase by as much as one power of ten at the temperature of liquid oxygen. Dyes with sensitizing and desensitizing properties exhibit a twofold low temperature effect: a decrease in the sensitivity in the sensitization bands and an increase in the blue sensitivity which has been lowered by de-sensitization at room temperature (see Fig. 15).

3. The spectral sensitivity distribution of sensitized emulsion layers containing supersensitizers does not usually decrease on cooling.

4. The relative decrease in sensitivity in the region of sensitization is in no way related to differences in the excitation energy of the sensitizers. According to *West*[213] cyanines and merocyanines of different chain lengths behave in the same way at low temperatures.

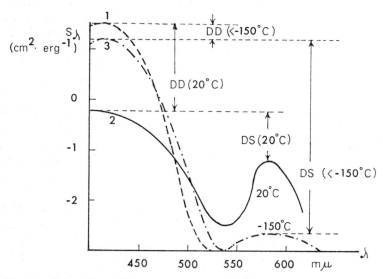

Fig. 15. Diagram of the effect of low temperature on the sensitizing and desensitizing action of a sensitizer. Spectral sensitivity distribution $S_\lambda = f(\lambda)$: 1. Unsensitized silver halide at 20°C. 2. Sensitized emulsion at 20°C. 3. Sensitized emulsion at < -150°C. DD: Degree of desensitization. DS: Degree of sensitization.

These findings make it evident that a sensitizing dye introduces into the elementary photographic process a reaction stage which is greatly dependent on temperature. Moreover the effects indicate the existence of a relation between sensitization and the opposing forces of desensitization.

3.1.2.4 *Spectral sensitization in relation to the thickness of the layer of dye.* The quantum yield of the sensitized photolysis of silver halide increases with decrease in the thickness of the layer of dye[40,163,215a] and reaches its highest value when the surface of the grains is covered with a monomolecular layer of dye molecules[198].

Therefore even if the sensitizing action of multimolecular layers is small in comparison with monomolecular ones, it is however certain that thick layers of dye are able to sensitize, and this holds for thicknesses of up to 1 μ.[215,216] This finding has an important bearing on the problem as to whether or not a radiationless transfer of energy is the cause of sensitization.

The decrease in sensitizing power which is shown by various sensitizing dyes, e.g. the cyanines[216,218] on changing over to multimolecular layers, and which is replaced by a desensitizing action, is remarkable. This "layer effect" by means of which a sensitizer assumes the behaviour of a desensitizer can likewise be regarded as a proof that a relation exists between the spectral sensitization and the desensitization of the photographic process.

3.1.2.5 *The arrangement of the Dye Molecules*. From measurements of adsorption isotherms which give the quantity of dye adsorbed per gramme of silver halide in relation to the concentration of the dye solution, it has already been possible to obtain a good idea of the arrangement of the dye molecules adsorbed to the grain surfaces.[219-222] In many cases the adsorption of the dye molecules takes place in accordance with the *Langmuir* adsorption isotherm. Not only does the latter indicate that the surface of a grain is covered with a monomolecular layer of dye molecules, but it also enables the number of molecules adsorbed to be calculated by taking readings of the saturation concentration from the horizontal section of the isotherm. From the surface requirements of the molecules and the area of the silver grain which can be covered, it is possible to derive the fact that the dyes are stacked one against the other with their long lateral edges orientated to the grain surface and that they are deposited plane to plane at a distance of about 3·6 Å from each other on the silver halide surface. X-ray investigations show moreover that the adsorbed dyes are inclined at about 60° to the grain surface.[223,224]

This description of the manner in which the dyes are deposited is also arrived at from the observation of J-bands in the absorption and reflection spectra, respectively, of the adsorbed dyes, for these bands are characteristic of the formation of polymeric dye aggregates which also occur in strongly concentrated solutions or in solutions containing strong electrolytes, and the structure of which has been studied by *Scheibe, et al.*[93-95,98,113,221-230]. This type of orientated dye adsorption is illustrated in Fig. 16.

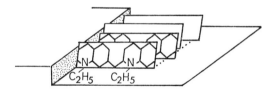

Fig. 16. The structure of adsorbed dye-aggregates.

Even if the described arrangement of the dye molecules is linked in many cases with the occurrence of a J-band—particularly in the pseudoisocyanines—nevertheless the absence of the characteristic band must not be taken to signify that a fundamentally different kind of deposition takes place[98,223], for the tendency of the molecules to stack themselves in layers with their planes towards one another in all probability represents a general principle of the arrangement of organic dyes during processes of concentration or crystallization, respectively, and it is often impossible to prove this principle simply on the basis of the superposition of the various groups of aggregates.[53]

The importance of the formation of aggregates in spectral sensitization might lie partly in the fact that the energy absorbed is able to migrate over many molecules before coming across a suitable place in the silver halide.[93] This energy migration effect is not however a prerequisite for the sensitization process since small aggregates of dye—even if they only give a small quantum yield—also sensitize.

In this connection it must be emphasized that the sensitizing action of dyes depends in every case on the adsorption layer being in a polymolecular state. The intermediate stages between the monomeric dye molecules and the polymeric J-aggregates, which can be detected in the carbocyanines by the presence of special D and H absorption bands,[53,94,96,198,231] possess less photochemical activity. For instance, the low sensitizing action of the methylates of a series of simple carbocyanines as compared with the corresponding ethylates, is due to the fact that the methylates are adsorbed mainly in the form of the H-states which possess little photochemical activity.[92,232,233] In addition to the simple *Langmuir* adsorption isotherms, in many cases complicated multi-stage forms of isotherm are often also observed during the adsorption of cyanines to silver halide microcrystals.[167,168,219–222,234–236] These forms are as a rule attributed to the fact that at low concentrations the dye molecules first cover the grain surface singly in a flat form, and at a critical concentration erect themselves in order to assume the characteristic aggregated form.[135] However, according to *Padday* and *Wickham*,[236a] the dye molecules which are adsorbed flat to the silver halide in the first monomolecular layer, retain their original orientation when multilayer adsorption takes place. The molecules which are taken up afterwards place themselves with their edges on the lower dye molecules and thus form a second layer in which the distances between the molecules corresponding to the J-states are small, being from 3·5–4 Å.

For further problems relating to the mechanism of the adsorption of sensitizing dyes to silver halides cf. *Boyer et al.*[111,237,238] Mention should also be made of the use of infra-red spectroscopy for investigating the interaction between a dye and a silver halide.[238,240]

3.1.2.6 *The Adsorption of Sensitizing Dyes in the Presence of Gelatin.* Since only an adsorbed sensitizer can transfer absorbed energy to a silver halide, the influence of various substances on the course of the sensitization reaction in photographic emulsions is of paramount importance. This might also be the reason why various dyes, although they can sensitize the photoconductivity of inorganic semiconductors, are incapable of producing any sensitizing effect in photographic emulsions. The compounds which are present in emulsions—gelatin and gelatin impurities, stabilizers, etc.[96]—are often more strongly adsorbed to the silver halide than is the dye, so that the latter cannot exert any sensitizing action. Thus a very characteristic feature of photographic sensitizers is their ability to expel completely, unwanted gelatin from the grain and to form a molecular layer around the grain. Other dyes on the other hand are dislodged to a greater or lesser extent by gelatin and thus lose their sensitizing power.[241] The type of gelatin also has some influence on this process.[242]

A rule which applies to symmetrical cyanine dyes is to the effect that the adsorption of dyes increases with the basicity of the heterocyclic nuclei and with the length of the methine chain up to the pentamethines.[243,244]

Stabilizers[245] and other emulsion additives also act in a similar way to gelatin because they expel the adsorbed dye to varying degrees and are thus inimical to the sensitization effect.[246] The sensitizing power of a dye is therefore only observed in photographic emulsions when the sensitizer is adsorbed directly to the silver halide grain independently of the presence of various additions.

This means that one of the most important prerequisites for the sensitization reaction is a close contact between the silver grain and the dye aggregate and one which cannot be influenced by additives.

3.1.2.7 *Displacements in Absorption.* Apart from the occurrence of *J*- or *H*-bands which are due to the formation of aggregates, the dyes undergo only slight displacements in absorption on being adsorbed. The absorption spectra of the relatively symmetrical but easily polarizable cyanines lie 10 to 40 mμ farther in the long wavelength region than those measured in solution, and in the case of the strongly polar merocyanines a displacement of 50 to 200 mμ in the sensitization spectrum is observed.[92,98,247,248] According to *Dörr*,[230] these displacements which occur when the dye is adsorbed are comparable with solvent effects and can be interpreted in a similar way.[53,249]

The intense, sharp bands of the *J*-aggregates which lie about 50 mμ farther in the long wavelength region than the molecular bands are displaced only very slightly on adsorption (50 mμ). These values show that only very insignificant changes of say from 0·03 to 0·15 eV take place in the energy levels (ground and excitation states) of sensitizing dyes on adsorption.

The values obtained with pure dye layers can therefore be taken as a basis when discussing the energy term position of dyes adsorbed to silver halide grains.

3.1.2.8 *The Fluorescence of Adsorbed Sensitizers.* Whereas fluorescing sensitizers show an intense emission of fluorescence on adsorption to magnesium oxide, silica gel and other non-photoconducting compounds, this fluorescent radiation is quenched on adsorption to AgBr and AgI (powder, emulsions, etc.).[36,40,250–255] This indicates that the excitation energy—either in the form of energy or as a charge—absorbed by the dye in silver halide/dye systems is transferred preferentially to the substratum undergoing sensitization, and is not given up by the dye with emission of fluorescence. The "luminescence of the 2nd kind"[256] which is observed in exceptional cases is not at variance with this conception.

Since the first excited singlet state from which the fluorescence originates lasts for about 10^{-9} seconds, it can therefore be concluded from the quenching of the fluorescence that the time taken for the transference of energy will be of an order of magnitude of less than 10^{-9} seconds.[253,257]

3.1.2.9 *The Participation of Triplet States.* The observation of the quenching of fluorescence makes the previously discussed assumption of the direct participation of triplet states in the sensitization reaction improbable.[258–260] This is because in the event of any co-operation between these states which are able to store the excitation energy, obtained by intersystem-crossing[260a] within 10^{-7}—10^{-10} sec. from the singlet state, for periods of the order of up to one second before releasing it without radiation or with the emission of phosphorescence, the complete extinction of luminescence would be incomprehensible.

A similar result has been arrived at from photoflash experiments in which, in contrast to the measurements made with non-adsorbed hydrocarbons and dyes,[261–263] no triplet-triplet absorption spectrum was manifested after intense irradiation.[251,253]

The fact that the quantum yield of triplet formation by intersystem-crossing from the first excited singlet to the triplet state in some cyanine dyes is less than 1 is also contradictory to a participation of triplet states in the sensitization reaction. For example the maximum quantum yields of triplet produced are 0.25 ± 0.05 for 1,1'-diethyl-6-iodo-2,2'-cyanine iodide, 0.08 ± 0.02 for 1,1'-diethyl-6-bromo-2,2'-cyanine iodide and 0.02 ± 0.01 for 1,1'-diethyl-2,2'-cyanine iodide.[263a] Since the quantum efficiencies in the dye-sensitzed region reach the order of 100 percent[208a] it is difficult to argue that triplets are formed as intermediate states in spectral sensitization.

3.1.2.10 *The Influence of Chemical Sensitization.* Chemical sensitization (e.g. by sulphur compounds) raises the sensitivity of a spectrally sensitized emulsion to the same extent as that of an undyed silver halide emulsion layer.[264–266] In other words, chemical sensitization by sulphur, etc. acts on the primary process of the production of the latent image and is not influenced by whether the latent image is formed by the light which is normally absorbed by the silver halide or by the light which is absorbed by the dye. Since sulphur-sensitization is equivalent to the adsorption of Ag_2S molecules[267] which may possibly act as hole traps and be able to lessen any recombination between the photoelectrons and the holes in the silver halide, the said observation appears to indicate that the spectral sensitization reaction is independent of the presence of the lattice defects contributed by chemical sensitization.

It must however be noted that numerous lattice defects are also present in pure silver bromide layers which are free from binding agent, as is proved by the anomalous red sensitivity of the layers of this kind discussed by *Eggert et al.*[42,49,57–59,268] The possible participation of lattice defects in the sensitization reaction—e.g. on regeneration of the dye[49,269]—therefore cannot be ruled out, in spite of *Carroll's*[265] observation regarding the missing influence of the chemical sensitization.

3.2 Special Sensitization Reactions

The scheme in Fig. 17 gives a general illustration of the reactions which can take place in sensitized photographic emulsion layers: desensitization, supersensitization and antisensitization. Not only is a knowledge of these reactions of practical importance, cf. Chap. 1, 2.1, but the processes of sensitization and

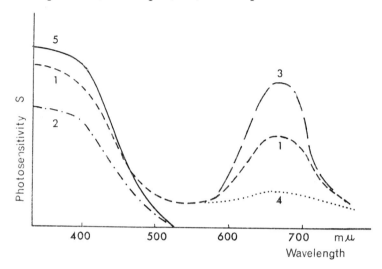

Fig. 17. Diagram of sensitization (1), desensitization (2), supersensitization (3), and antisensitization (4). The normal sensitivity of unsensitized AgBr emulsions (5).

supersensitization are also of particular interest in explaining the mechanism of spectral sensitization, since these effects are most probably closely connected with one another, and can be classified in a general scheme of reactions.

3.2.1 DESENSITIZATION. By desensitization is understood the decrease in the sensitivity of unsensitized silver halide emulsions brought about by dyes.[41,42,92,198,218,135,270-273]

In many cases this effect can, of course, be caused simply by an optical filtering action on the part of the added dye. For the most part however a special reaction might take place and inhibit the production of the latent image nuclei in the photographic emulsion layer on exposure of the emulsion. Apart from the prerequisites which first decree whether or not a dye can be used as a desensitizer in photographic practice and thus enable exposed silver halide emulsion layers to be developed in comparatively bright light (e.g. solubility, power of resistance at a high pH and in the presence of sulphite, strong power of adsorption to silver halide[135]), the following mechanisms governing the action of desensitization are of importance:

1. Not only is desensitization achieved by means of typical desensitizing dyes such as Phenosafranine, Methylene blue, etc.,[274] but it is also observed with sensitizers. It appears that sensitizers (cyanines, etc.) can assume desensitizing properties when their concentration is raised, i.e. when they exist in the form of a multimolecular layer adsorbed to a silver halide grain.[45,217,218,275] After having reached a maximum sensitivity the integral sensitivity of sensitized emulsion layers therefore decreases on raising the concentration of the dye because the increase in speed achieved by sensitization is compensated by the desensitization, particularly in the case of high dye concentrations.[44]

The double function of sensitization and desensitization exhibited by a dye might be of the greatest importance in explaining the mechanism of sensitization, for the validity of any theory of sensitization can be tested by its ability to provide an explanation of this reciprocal effect. The alternation between sensitization and desensitization which is dependent on concentration is of practical importance in ensuring that the various dyes are present in their respective optimum concentrations for obtaining a sensitization effect.[42,276]

2. In dyes of similar constitution distinct relations exist between sensitization and desensitization, respectively, and chemical constitution. According to *Riester et al.*,[36,130,217,277] a sensitizer can be changed into a desensitizer by replacing the methine group —CH= in polymethine dyes by —N=, or by attaching a —CN= group to a methine group. This conversion is accompanied in a remarkable way by a change in absorption which is characterized by the fact that the desensitizing dye possesses an absorption band situated 100 mμ farther towards the long wavelengths.

As examples we shall cite again the dyes Acridine yellow and Safranine which differ in the nature of their bridging atoms. Acridine yellow [47] the ring of which is closed by a carbon atom and which has a λ_{max} of 456 mμ, represents a sensitizer, whereas Safranine [48], on account of its N bridge, has a desensitizing action and possesses a long wavelength absorption maximum at 539 mμ. Cf. further examples in[53,130,217] and in Chap. II, section 2.3.

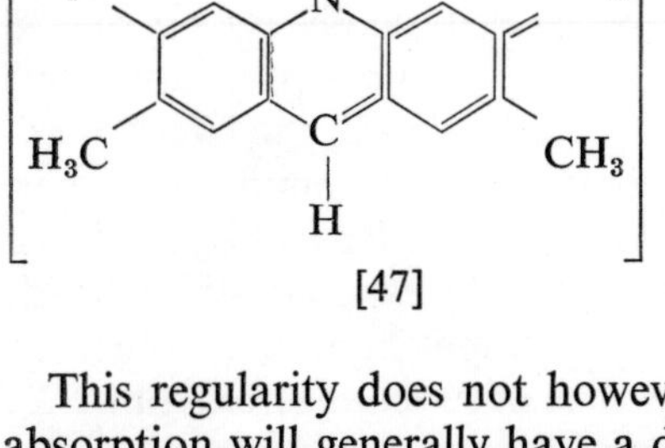

[47] 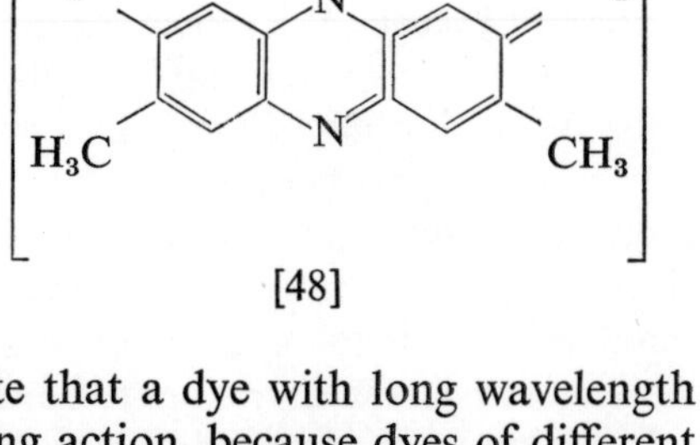 [48]

This regularity does not however indicate that a dye with long wavelength absorption will generally have a desensitizing action, because dyes of different constitution, but likewise showing a characteristic long wavelength absorption are entirely capable of acting as sensitizers. In the case of dyes of equal chain length the displacements in absorption are due to the fact that the introduction into the middle of the chain of an atom of greater electron affinity than that possessed by the carbon atom, influences either the ground state or the excitation state depending on the length of the chain,[217,278,279] and therefore the observed regularity indicates that a connection exists between the position of the excitation state (and of the first unoccupied level, respectively) and the desensitizing action. A lowering of the ground state and of the highest occupied level respectively in the dye does indeed result in a short wavelength displacement, whereas a depression of the excitation state brings about the observed long wavelength displacement.

From this relation between constitution and sensitizing and desensitizing action respectively, *Scheibe* and *Dörr*[280,281] deduce that:

1. a dye sensitizes when the energy level of the 1st excitation state lies about 0·1 eV above the conduction band of the silver halide;

2. a dye desensitizes when the excitation state lies 0·2 to 0·5 eV below the conduction band of the silver halide. According to this hypothesis photoelectrons are trapped by the dye, and this action would therefore represent the actual cause of desensitization as discussed by say *Gross*.[282] Thus the simultaneous sensitizing and desensitizing action of dyes would accordingly be explained by the large degree of coincidence between the excitation level in the dyes and the conduction band of the silver halide.

Without going any further into the problems raised by this electron-acceptor action—cf. also the discussions on whether desensitization may possibly be caused by a retardation of the development[45,283-286]—it should also be pointed out that the experimental results obtained by *Hada, Tamura et al.*[217,287,287a] fit in well with the specified hypothesis. From the experiments it follows namely that strong desensitizers have a lower (up to 0·7 eV) first vacant energy level than do weak desensitizers, or that the redox potential of desensitizing dyes is higher than that of powerful sensitizers.[288] Measurements of the cathodic polarographic half-stage potential $E_{\frac{1}{2}}$, which can be taken as a measure of the redox potential and of the relative position of the lowest unoccupied level in a dye,[41,217,289] show that the variations in the photographic action of the thiacyanines are likewise caused by a drop in the lowest vacant level on changing over from a thiacarbocyanine to a thiadicarbocyanine and to a thiatricarbocyanine. On the basis of these measurements the excitation level of a thiatricarbocyanine which acts as a desensitizer is found to be situated 0·72 eV below the corresponding level of a thiacyanine which acts as a sensitizer. The long wavelength displacement of the maximum absorption corresponding to this lowering in the lowest vacant level—cf. also 290, 291—on changing over from a thiacyanine to a thiatricarbocyanine[213] thus proves that the ground state of these dyes which are similar in constitution is in agreement. Similar relations are likely to exist within other series of dyes, e.g. the merocyanines.

In this connection we would also draw attention to the measurements which were made by *Meiklyar, et al.*,[292] and which show that the reduction by Phenosafranine of the spectral sensitivity of silver halide emulsion layers which have been sensitized by 3,3'-diethylthiacarbocyanines, decreases on lengthening the polymethine chain in the sensitizer.

3. The sensitizing and desensitizing actions respectively of dyes also depend on the basic photographic material. For instance azacyanine is a sensitizer for silver chloride but not for silver bromide.[293] In general, silver iodide can be desensitized more easily than silver bromide or silver chloride,[294] and dyes which have a sensitizing action on AgCl can therefore be powerful desensitizers of silver iodide emulsion layers.

It is natural to attribute this effect to differences in the positions of the energy levels in the silver halides. *Scheibe*[280,281] however points out that the lattice defects on the surface of the crystal may have a decisive influence in helping to bring about this anomalous behaviour.

4. Just as with the sensitizing action, strong adsorption of the dye to the silver halide also decides the way in which the reaction proceeds in desensitization.[295] For instance the large difference between the desensitizing action of nitrobenzimidazole and nitrochlorobenzotriazole both of which, according to their polarographically estimated redox potentials, should be good desensitizers, might be due to an anomaly in their adsorption behaviour. Measurements of the infra-red spectrum of a dye adsorbed to AgCl show, namely that in comparison with nitrobenzimidazole, nitrochlorobenzotriazole which has a desensitizing action is strongly adsorbed.[289]

5. The desensitizing action decreases with fall in temperature. *Frieser et al.*[109] observed that with dyes which possess both sensitizing and desensitizing properties, the sensitizing and desensitizing actions disappear completely at $-176°$. With strong desensitizers such as Phenosafranine or Pinakryptol green the effect of temperature is weaker, but nevertheless it is clearly perceptible.

6. Whereas the photoconductivity of silver halides can be spectrally sensitized in a similar manner to their photographic sensitivity by the use of dyes,[296,297] a corresponding desensitization effect has not yet been observed. The photoconductivity of a silver halide which is observable in the normal region of absorption of AgBr (i.e. at $\lambda < 500$ mμ) is not diminished by desensitizing dyes. On the contrary, on irradiation in the absorption region of the desensitizer, the photoconduction of AgBr even appears to increase in some cases, i.e. it can be sensitized by the desensitizer.[36,298]

This effect is of particular importance in explaining photographic sensitization, since it casts great doubt on some of the theories which have been discussed previously.[198]

Finally we would draw attention to the problem of fog formation[299] as well as to the fact that the dependence on wavelength of the sensitivity of X-ray emulsions to X- and gamma-rays can be influenced by desensitizing dyes.[300]

3.2.2 SUPERSENSITIZATION. By supersensitization is understood the large increase in sensitivity in the region of spectral sensitization produced by the addition of certain organic substances—the absorption of which is so small as to be not worth mentioning—to silver halide emulsions[36,38,40,225,296,297] and to photoconducting layers.

According to *Scheibe*[95,297] two types of this effect can be distinguished:

1. Supersensitization of the 1st kind which is due to the formation of polymers of the sensitizing dye by the action of auxiliary polymerizing agents. This effect is characterized by a change in the dye absorption, as the aggregates formed exhibit characteristic absorption bands (*J*-bands, etc.). The supersensitizers of the 1st kind which are usually added in excess, include aromatic heterocyclic nitrogen bases and their derivatives as well as esters, amides, ketones, nitriles, isocyclic aromatic compounds, etc.

2. Supersensitization of the 2nd kind which is released by substances which are added in small quantities to an emulsion and which have no influence on the sensitizers which are deposited on the silver halide. This type of reaction is usually regarded as true supersensitization.

On account of the practical importance of increasing the spectral sensitivity of emulsions, this effect which was first observed by *Bloch* and *Renwick*[301] in the year 1920 is described predominantly in the patent literature, and for this reason the reader is referred to the information given in the reviews by *Mudrovcic, Scheibe et al.*[232,297,302] as well as to a few recent patents.[303–308] The following statements can be made regarding the relation between the constitution and the supersensitization of a compound:[297] interaction between the supersensitizer and the dye is essential for supersensitization, and for this to take place the compound must exert about the same forces in relation to adsorption and aggregation as those exerted by the sensitizing dye. Since these forces are due to an extended π-electron system, substances, which in theory are similar in nature to sensitizers, such as those discussed at length in Chap. II (section 3) make suitable supersensitizers. Therefore supersensitizers of the amidinium-ion, the carboxyl-ion and the amide type as well as their derivatives are to be found. For the supersensitization of 1,1'-diethyl-2,2'-cyanine chloride [49] (*M*-band: 546 mμ; *J*-band: 578 mμ)

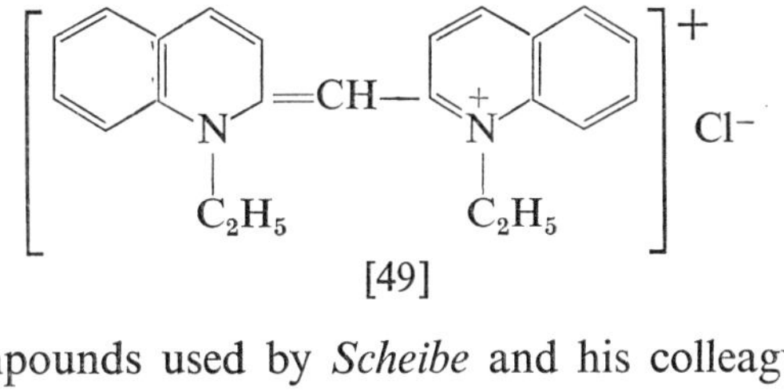

$$\left[\text{Quinoline-N}(C_2H_5)=CH-\text{Quinoline-N}^+(C_2H_5) \right]^+ Cl^-$$

[49]

the following compounds used by *Scheibe* and his colleagues[297] are given as examples.

1. Bases which are derived from quinoline, benzothiazole, pyridine and other

heterocyclic compounds of the type 2-(*p*-dimethylaminostyryl)-quinoline [50] or 2-(*p*-dimethylaminostyryl)-benzothiazole [51]. Compound [51] for example, raises the green sensitivity of a weakly sensitizing quino-pseudocyanine by a factor of four.

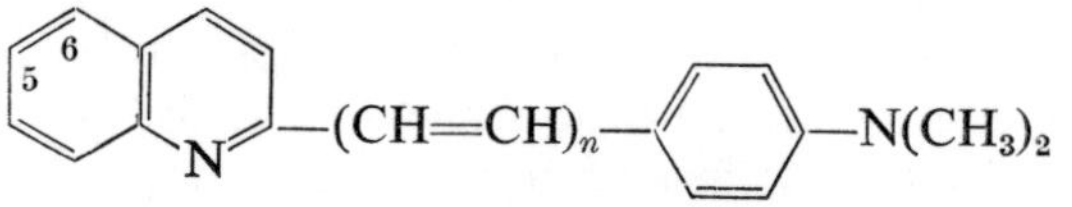

[quinoline]—(CH=CH)$_n$—[phenyl]—N(CH$_3$)$_2$

[50]

[benzothiazole]C—(CH=CH)$_n$—[phenyl]—N(CH$_3$)$_2$

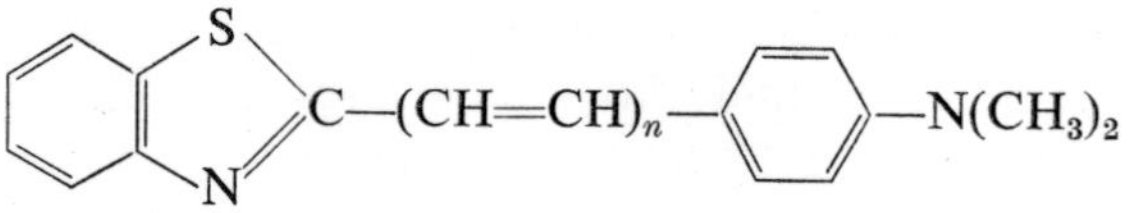

[51]

The nature of the alkyl substituents attached to the amino group is without influence in the substances named: substitution by halogens in the 6-position improves supersensitization, and anellation in the 5,6-positions has a similar effect. Quinoline derivatives are less active than the analogous compounds of other heterocyclic compounds.

2. Cationic dyes, which are formed by quarternizing the said bases, exhibit little or no supersensitization in comparison with the bases. There is, however, one series of these dyes which show good effects like the supersensitizers corresponding to the hemicyanine [52] and to the cyanine [53].

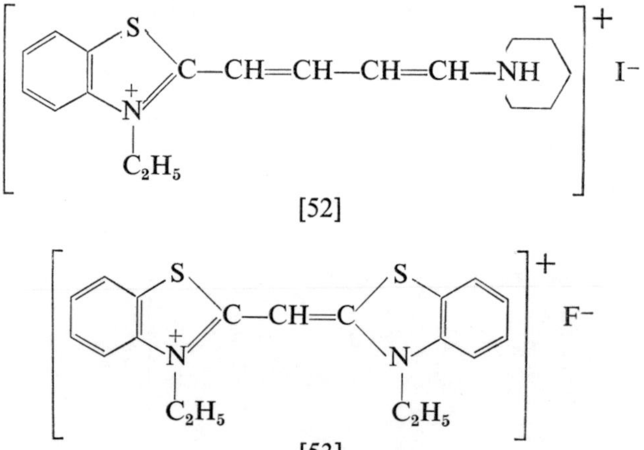

$$\left[\text{[benzothiazole, N}^+\text{-C}_2\text{H}_5\text{]C—CH=CH—CH=CH—NH[piperidine]} \right]^+ \quad \text{I}^-$$

[52]

$$\left[\text{[benzothiazole, N}^+\text{-C}_2\text{H}_5\text{]C—CH=C[benzothiazole, N-C}_2\text{H}_5\text{]} \right]^+ \quad \text{F}^-$$

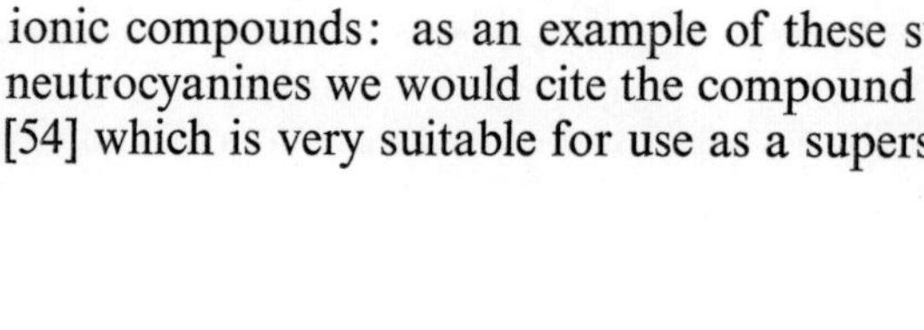

[53]

3. Amphoteric ionic compounds: as an example of these substances which are similar to the neutrocyanines we would cite the compound derived from an *o*-oxy-styryl base [54] which is very suitable for use as a supersensitizer.

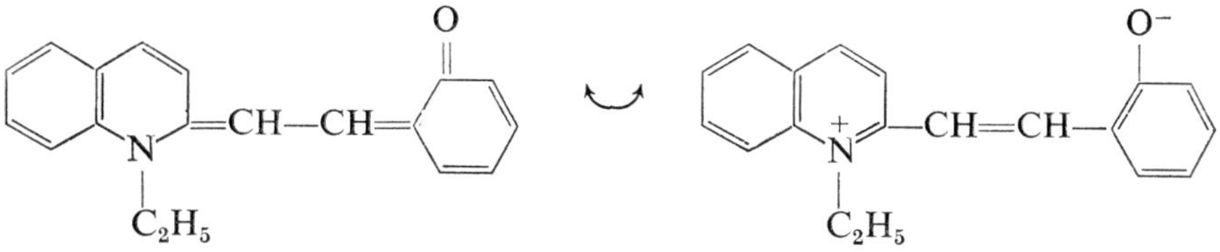

[54]

4. Anionic compounds produce no supersensitization in emulsions which have been sensitized by dye [49], possibly because the carboxyl ions are only very slightly adsorbed to the silver halide, and thus interaction with the sensitizing dye no longer takes place.

In summarizing it may be said that the examples given show that compounds which are similar and dissimilar in structure to the sensitizing dye are equally able to exert a supersensitizing action on the sensitizer. Even the symmetry of the molecules does not play a decisive part. That is to say, in addition to certain structural influences [e.g. —N(CH$_3$)$_2$ must be present instead of —NH$_2$ in [50] or [51]; in [52] supersensitization is only observed from a tetramethine chain onwards] it is above all the additional factors such as the interaction with the dye or the adsorption to the silver halide which play a decisive part in producing the effect. It is therefore not yet possible at the present time to make general predictions of the suitability or otherwise of compounds for supersensitizing in relation to certain sensitizers.

The following facts are worthy of note in regard to the mechanism of the effect.

1. Supersensitization occurs not only with dye aggregates but it also occurs with monomeric adsorbed dye molecules.[297]

2. The photoconductivity of sensitized silver halide emulsions is intensified by a supersensitizer in a similar manner to the photographic sensitivity of these emulsion layers.[36]

3. Supersensitizers quench any fluorescence which may arise in dye aggregates which have been influenced photographically by them, both when the dyes are in solution and when they are in the adsorbed state.

4. Supersensitizers already take effect when they are present in very small quantities, i.e. when the ratio of supersensitizer to sensitizer is small.

How well these different effects can be fitted in their proper sequence into the mechanism of sensitization and supersensitization will be discussed in Chap. VII; cf. also *Gross, Scheibe et al.*[79a,297]

3.2.3 ANTISENSITIZATION. Antisensitization is the term used to describe the decrease in spectral sensitivity in the region of dye-sensitization (see Chap. I, section 2.1). Antisensitizers which are already active in low concentrations are non-planar dyes, the absorption bands of which largely correspond with that of the sensitizer. This effect only occurs when the absorption of the antisensitizer does not lie more than 50 mμ farther in the short wavelengths than that of the sensitizer.

As an example we shall cite the antisensitization of 1,1'-diethyl-2,2'-cyanine iodide [1] by 1,1',3,3'-tetramethyl-2,2'-cyanine [55], which, according to *West* and *Carroll*,[38] decreases the sensitivity in the region of sensitization from $S_{10/i} = 16\cdot5$ to $S_{10/i} = 4\cdot5$, with only a slight influence on the blue sensitivity. As in the case of supersensitization, the effect is already perceptible with a small antisensitizer/sensitizer ratio (1 : 10).

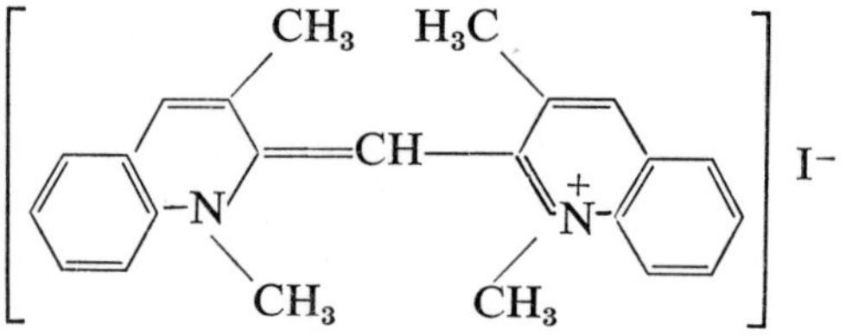

[55]

It is significant that in the case of non-planar dyes of the type 1,1',3,3'-tetramethyl-2,2-cyanine [55], the antisensitizing action in relation to sensitizers which absorb more than 50 mμ farther in the long wavelengths, can be changed into supersensitization. Dye [55], with a λ_{max} of 530 mμ is a strong antisensitizer for aggregated thia-2'-cyanines and 2,2'-cyanines, the *J*-bands of which lie at a shorter wavelength than 580 mμ. 1,1'-diethyl-5,6,5',6'-dibenzo-2,2'-cyanine [56] which is characterized by a *J*-band at 600 mμ, is however supersensitized by dye [55].[38]

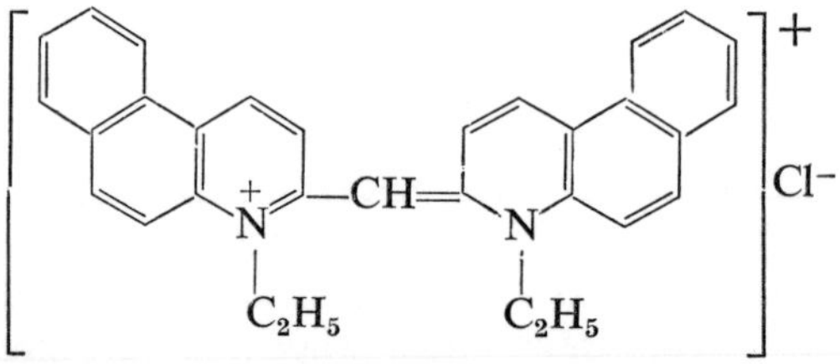

[56]

A fact which might probably be deduced from this observation is to the effect that the mechanisms of antisensitization and supersensitization which act in opposite directions are not comparable with each other. The analogy to fluorescence quenching (cf. Chap. I, 2.1./3, equation 5) as well as the overlapping condition of the absorption bands makes it natural to assume that the transfer of energy is a radiationless process in keeping with the one proposed by *Förster* for antisensitization. According to this an antisensitizer which is built into a dye aggregate would first accept the energy absorbed by the aggregate without the emission of radiation. After this the non-planarity of the antisensitizer is then responsible for the preferential conversion into and consumption of the electronic excitation energy in torsional vibrations, so that spectral sensitization of the silver halide does not take place.[297]

It should also be noted that the dye-sensitized photoconductivity of silver halide emulsions also decreases when antisensitizers are added. Thus an antisensitizer acts directly in the primary process of sensitization.

References

[198] WOLFF, H., *Fortschr. chem. Forsch.*, 3/3, 5031 (1955)).

[199] CARROLL, B. H., *Photogr. Sci. Eng.*, **5**, 65 (1961).

[200] BOURDON, J. and DURANTE, M., *J. Chim. phys.*, **60**, 863 (1963).

[201] DICKINSON, O., *J. photogr. Sci.*, **2**, 50 (1954).

[202] BAGDASAR'YAN, KH. S., *Acta physicochim. USSR*, **9**, 205 (1938).

[203] SHEPPARD, S. E., LAMBERT, R. H. and WALKER, R. D., *J. chem. Phys.*, **7**, 426 (1939).

[204] EGGERT, J. and NODDACK, W., *Z. Phys.*, **31**, 922 (1925).

[205] EGGERT J. and NODDACK, W., *Z. Phys.*, **34**, 918 (1925).

[206] WILCKE, E., *Dissertation*, (*Thesis*), Universität Freiburg i. Br. 1941.

[207] SOLOV'EV, S. M. and YAMPOL'SKII, P. A., *Zh. fiz. Khim.*, **21**, 907 (1947).

[208] TOLLERT, H., *Z. phys. Chem.*, **140A**, 355 (1929).

[208a] SAGE, J. P., *J.phys. Chem. Solids*, 26, 1245 (1965).

[209] LESZYNSKI, W., *Z. wiss Photogr.*, **24**, 261 (1926).

[210] EGGERT, J. and NODDACK, W., *Naturwissenschaften*, **15**, 57 (1927).

[211] EGGERT, J., MEIDINGER, W. and AHRENS, H., *Helv. chim. acta*, **31**, 1163 (1948).

[212] MECKE, R. and SEMERANO, G., *Z. Electrochem.*, **40**, 511 (1934).

[213] WEST, W., *Photogr. Sci. Eng.*, **6**, 92 (1962).

[214] EVANS, C. H., *J. opt. Soc. Amer.*, **32**, 214 (1942).

[215] MEIER, H. and ALBRECHT, W., *Ber. Bunsenges. phys. Chem.*, **67**, 838 (1963); **68**, 64 (1964).

[215a] ZUCKERMAN, B., *Photogr. Sci. Eng.*, **11**, 156 (1967).

[216] NELSON, R. C., *J. opt. Soc. Amer.*, **46**, 13 (1956).

[217] TAMURA, M. and HADA, H., *Scientific Photography*, Proc. Int. Colloq., Liége 1959, edited by H. SAUVENIER, p. 572, Pergamon Press (Oxford) 1962.

[218] HEISENBERG, E., *Veröffentlichungen des wissenschaftlichen Zentral-Laboratoriums der photographischen Abteilung-AGFA*, (*Publications of the Photographic Department of the AGFA Central Scientific Laboratory*), Vol. III, p. 115, Hirzel (Leipzig) 1933.

[219] SHEPPARD, S. E. and CROUCH, H., *J. phys. Chem.*, **32**, 751 (1928).

[220] SHEPPARD, S. E., LAMBERT, R. H. and KEENAN, R. L., *J. phys. Chem.*, **36**, 174 (1932).

[221] SHEPPARD, S. E., LAMBERT, R. H. and WALKER, R. D., *J. chem. Phys.*, **7**, 265 (1939).

[222] WEST, W., CARROLL, B. H. and WITHCOMB, D. L., *J. phys. Chem.*, **56**, 1054 (1952).

[223] HOPPE, W., *Kolloidzschr.*, **109**, 21 and 27 (1944).

[224] SONOIKE, S. and OKABE, H., *J. appl. Phys. Japan*, **26**, 392 (1957); *Sci. Industr. photogr.*, **29** (2), 137 (1958).

[225] FÖRSTER, TH., *Naturwissenschaften*, **33**, 166 (1946).

[226] SCHEIBE, G., KANDLER, L. and ECKER, H., *Naturwissenschaften*, **25**, 75 (1937).

[227] SHEPPARD, S. E., *Rev. mod. Phys.*, **14**, 303 (1942).

[228] NATANSON, S. V., *Dokl. Akad. Nauk. SSSR*, **106**, 497 (1956).

[229] WEST, W. and SAUNDERS, V. I., *Wissenschaftliche Photographie* (*Scientific Photography*) (Int. Konf. Köln, 1956) Edited by O. HELWICH, p. 48 (Darmstadt) 1958.

[230] DÖRR, F., *Photographic Science*, Symposium Zürich 1961, edited by W. F. BERG, p. 81, Focal Press (London) 1963.

[231] SCHEIBE, G., SCHÖNTAG, A. and KATHEDER, F., *Naturwissenschaften*, **29**, 499 (1939).

[232] LEVKOEV, I. I. and LIFSHITS, E. B., *Zh. nauch. prikl. Fotogr. Kinem.*, **3**, 419 (1958).

[233] LEVKOEV, I. I., LIFSCHITS, E. B. and NATANSON, S. V., *Scientific Photography*, Proc. Int. Colloq. Liége 1959, edited by H. SAUVENIER, p. 440, Pergamon Press (Oxford) 1962.
[234] NATANSON, S. V., *Scientific Photography*, Proc. Int. Colloq. Liége, 1959, edited by H. SAUVENIER, p. 457, Pergamon Press (Oxford) 1962.
[235] DAVY, E. P., *Trans. Faraday Soc.*, **36**, 323 (1940).
[236] PADDAY, J. F., *Trans. Faraday Soc.*, **60**, 1325 (1964).
[236a] PADDAY, J. F. and WICKHAM, R. S., *Trans. Faraday Soc.*, **62**, 1283 (1966).
[237] SOLOV'EV, S. M. and RODIONOVA, N. I., *Zh. nauch. prikl. Fotogr. Kinem.*, **7**, 81 (1962).
[238] KIRILLOV, E. A. and RAKITYANSKAYA, O. F., *Zh. nauch. prikl. Fotogr. Kinem.*, **10**, 28 (1965).
[239] BOYER, S., *C.R. hebd. Séances Acad. Sci.*, **258**, 178 (1964).
[240] BOYER, S., MALINGREY, B. and PRETESEILLE, M. C., *Sci. Ind. photogr.*, **36**, (2), 217 (1965).
[241] JAMES, T. H. and VANSELOW, W., *J. Amer. chem. Soc.*, **73**, 5617 (1951).
[242] SATO SHUNI and BABA SHIGEJI, *Nippon Shashin Gakkai Kaishi*, **28**, 70 (1965).
[243] WEST, W., CARROLL, B. H. and WITHCOMB, D. L., *J. photogr. Sci.*, **1**, 145 (1953).
[244] KLEIN, E. and MOLL, F., *Photogr. Sci. Eng.*, **3**, 232 (1959).
[245] MEYER, K. and POLENZ, H. J., *Z. wiss Photogr.*, **54**, 81 (1960).
[246] KLEIN, E. and MOLL, F., *Scientific Photography*, Proc. Int. Colloq. Liége 1959, edited by H. SAUVENIER, p. 470, Pergamon Press (Oxford) 1962.
[247] DEICHMEISTER, M. V., LEVKOEV, I. I., LIFSCHITS, E. G. and NATANSON, S. V., *Dokl. Akad. Nauk. SSSR*, **93**, 1057 (1953).
[248] WEST, W., CARROLL, B. H. and WITHCOMB, D. L., *Ann. New York Acad. Sci.*, **58**, 893 (1954).
[249] WELLER, A., *Disc. Faraday Soc.*, **27**, 28 (1959).
[250] HORWITZ, L. and FRIEDMAN, J., *J. opt. Soc. Amer.*, **45**, 798 (1955).
[251] TERENIN, A. and AKIMOV, I., *Scientific Photography*, Proc. Int. Colloq. Liége 1959, edited by H. SAUVENIER, p. 532, Pergamon Press (Oxford) 1962.
[252] DÖRR, F. and SCHEIBE, G., *Z. wiss Photogr.*, **55**, 133 (1961).
[253] WEST, W., *Scientific Photography*, Proc. Int. Colloq. Liége 1959, edited by H. SAUVENIER, p. 557, Pergamon Press (Oxford) 1962.
[254] AKIMOV, I. A., *Zh. nauch. prikl. Fotogr. Kinem.*, **4**, 64 (1959).
[255] MEIDINGER, W., *Phys. Z.*, **40**, 517 (1939); **41**, 277 (1940).
[256] KICIAK, K. and BASINSKI, A., *Roczniki Chem.*, **39**, 1847 (1966).
[257] SAZEPIN, I., *Zh. nauch. prikl. Fotogr. Kinem.*, **5**, 60 (1960).
[258] SIMPSON, W. T., *J. chem. Phys.*, **15**, 414 (1947).
[259] CLIMENTI, E. and KASHA, M., *J. chem. Phys.*, **26**, 956 (1957).
[260] UMANO, S., *Bull. Soc. sci. photogr. Japan*, **2**, 7 (1957).
[260a] TURRO, N. J., *Molecular photochemistry*, p. 64, W. A. Benjamin (New York, Amsterdam) 1965.
[261] PORTER, G. and WINDSOR, M. M., *Disc. Faraday Soc.*, **17**, 178 (1954).
[262] PORTER, G. and WRIGHT, F. G., *Trans. Faraday Soc.*, **51**, 1205 (1955).
[263] LIVINGSTON, B., PORTER, G. and WINDSOR, M., *Nature, Lond.*, **173**, 485 (1954).
[263a] BUETTNER, A. V., *J. chem. Phys.*, **46**, 1398, 1967.
[264] CARROLL, B. H. and HUBBARD, D., *Bur. Stand. J. Res.*, **9**, 529 (1932).
[265] CARROLL, B. H., WILLIAM, E. A. and HENRICKSON, R. B., *Photogr. Sci. Eng.*, **5**, 230 (1961).
[266] EGGERS, J., KLEIN, E. and MATEJEC, R., *Angew. Chem.*, **69**, 291 (1957).
[267] MITCHELL, J. W., *Die photographische Empfindlichkeit* (*Photographic Sensitivity*) (*Photogr. Korr.*, 1st special issue), O. Helwich (Darmstadt) 1957.
[268] CHIBISOV, K. V. and CHELTSOV, V. S., *Trudy Vses Nauchno-issled. Kinofotoinst.*, **1**, 128 (1932).
[269] NODDACK, W. and MEIER, H., *Z. Elektrochem.*, **63**, 971 (1959).
[270] CARROLL, B. H., *Photographic Science*, Symposium Zürich 1961, edited by W. F. BERG, pp. 90 and 95, Focal Press (London) 1963.
[271] LÜPPO-CRAMER, H., *Z. wiss Photogr.*, **31**, 329 (1933); **36**, 8 (1937).
[272] BLAU, M. and WAMBACHER, H., *Z. wiss Photogr.*, **33**, 191 (1934); *Nature, Lond.*, **134**, 538 (1934); *Photogr. Korr.*, **72**, 108 (1936).
[273] BOKINIK, YA. A. and SMIRNOVA, V. A., *Kino foto-khim Prom.*, **3**, 51 (1937).
[274] Fr. P. 1,408,108 (Cl. G 03c), H. B. COWDEN (to Kodak-Pathé) Aug. 6, (1965).
[275] LÜPPO-CRAMER, H., *Photogr. Korr.*, **63**, 329 (1927).

[276] KÖNIG, E., *Das Arbeiten mit farbenempfindlichen Platten* (*Methods of Working with Colour-sensitive Plates*), p. 46, G. Schmitt (Berlin) 1909.
[277] HAUTOT, A. and SAUVENIER, H., *Bull. Soc. r Sci. Liége*, **21**, 79 and 514 (1952); *Sci. Ind. Photogr.*, **23** (2), 137 (1952).
[278] FÖRSTER, TH., *Z. phys. Chem.*, **B48**, 12 (1940).
[279] DEWAR, M. J. S., *J. chem. Soc.*, 2329 (1950).
[280] SCHEIBE, G. and DÖRR, F., *Scientific Photography*, Proc. Int. Colloq. Liége 1959, edited by H. SAUVENIER, p. 512, Pergamon Press (Oxford) 1962.
[281] SCHEIBE, G. and DÖRR, F., *Sitzber. Math. Naturw. Kl. Bayer Akad. Wiss.*, München, p. 183, 1959.
[282] GROSS, L. G., *Trudy Vses Nauchno-issled. Kinofotoinst.*, **37**, 64 (1960).
[283] BLAKE, R. K., *Photogr. Sci. Eng.*, **9**, 91 (1965).
[284] JAMES, T. H. and VANSELOW, W., *Photogr. Sci. Eng.*, **6**, 183 (1955).
[285] BORIN, A. V., *Zh. nauch prikl Fotogr. Kinem.*, **9**, 215 (1964)
[286] KHEINMAN, A. S., DONATOVA, V. P. and BOCHAROVA, L. N., *Zh. nauch prikl Fotogr. Kinem.*, **11**, 61 (1966).
[287] HADA, H. and TAMURA, M., *Bull. Soc. sci. photogr.*, *Japan*, **7**, 1 (1957).
[287a] TAMURA, M. and HADA, H., *photogr. Sci. Eng.*, **11**, 82 (1967).
[288] SHEPPARD, S. E., LAMBERT, R. H. and WALKER, R. D., *J. phys. Chem.*, **50**, 210 (1946).
[289] REUTENAUER, G., *Photographic Science*, Symposium Zürich 1961, edited by W. F. BERG, pp. 24 and 95, Focal Press (London) 1963.
[290] MACCOLL, A., *Nature, Lond.*, **163**, 178 (1949).
[291] LYONS, L. E., *Nature, Lond.*, **166**, 193 (1950).
[292] MEIKLYAR, P. V., MIRMIL'STEINEBERMAN, M. D. and SADYKOVA, A. A., *Zh. nauch prikl Fotogr. Kinem.*, **10**, 401 (1965).
[293] HAMER, F. M. and FISHER, N. I. *J. chem. Soc.*, 90 (1937).
[294] LÜPPO-CRAMER, H., *Photogr. Industr.*, **31**, 432 (1933); *Photogr. Korr.*, **68**, 197 (1932).
[295] GOROKHOVSKII, YU. N., LEVIN, YA. A., KISELEVA, I. P. and GALIMOVA, A. M., *Zh. nauch prikl Fotogr. Kinem.*, **8**, 205 (1963).
[296] WEST, W. and CARROLL, B. H., *J. phys. Chem.*, **57**, 797 (1953).
[297] BRÜNNER, R., OBERTH, A. E., PICK, G. and SCHEIBE, G., *Z. Elektrochem.*, **62**, 132 and 146 (1958).
[298] CARROLL, B. H. and WEST, W., *Fundamental Mechanisms of Photographic Sensitivity*, Proceedings of a Symposium held at the University of Bristol in 1950, Butterworth, 1951.
[299] ZYUSKIN, N. M., *Zh. nauch. prikl. Photogr. Kinem.*, **9**, 11 (1964).
[300] KLEIN, D., *Photogr. Korr.*, **98**, 67 (1962).
[301] BLOCH, O. and RENWICK, F. F., *Photogr. J.*, **60**, 145 (1920).
[302] MUDROVCIC, M., *Sci. Ind. photogr.*, **24** (2), 47 (1953).
[303] Fr. P. 1,401,627 (Cl. G 03c), E. JONES and N. W. KALENDA (to Kodak-Pathé) June 4 (1965).
[304] Belg. P. 659,853, J. A. SCHWAN and J. E. JONES, June 16 (1965).
[305] Ger. (East) P., 39,717 (Cl. G 03c), J. BRUNKEN, E. J. POPPE and W. SCHINDLER, Aug. 5 (1965).
[306] Belg. P. 660,948, O RIESTER (to Agfa A.G.), Sept 13 (1965).
[307] Ilford Ltd., *Meth. Appl.*, 6,409,727 (Cl. G 03c), March 1, 1965.
[308] Ilford Ltd., *Meth. Appl.*, 6,504,708 (Cl. G 03c), Oct. 21, 1965.

IV. THE SPECTRAL SENSITIZATION OF PHOTOCONDUCTIVITY

4.1 *The Photoconductivity of Spectrally Sensitized Silver Halides*

4.1.1 THE MECHANISM OF THE FORMATION OF THE LATENT IMAGE. In attempting to solve the problems involved in spectral sensitization it is essential that the actual photographic process shall not be altered in any way by the process of sensitization. Here again a latent image might be formed in accordance with the Gurney-Mott theory [309,310-312] via an electronic and an ionic process.

In this connection it should be pointed out that this theory of the elementary photographic process is in accordance with the results of numerous investigations which include the testing of the photochemical equivalency law and estimations of the quantum yield, respectively,[203-205,313-315] the proof given by *Berg, Marriage* and *Stevens*[316,405] that the distribution of the latent image is independent of the wavelength of the light absorbed, measurements of the spreading of the internal latent image to the surface image under certain conditions,[317a-c] and the quantum sensitivity.[318,319] Mention must also be made of measurements of electron spin resonance,[320,321] and the luminescence of silver halides,[322-325] electron microscope investigations of the silver formed photolytically in halide grains,[326,327] X-ray interference measurements, etc.[328-331] Above all, measurements of the dark and photoconductivity of silver halides are of special importance,[332-337] as well as the experiments on the print-out effect using the voltage pulse technique introduced by *Haynes* and *Shockley*.[338] As long ago as 1930, on the basis of the agreement between the photographic sensitivity and the photoconductance of silver bromide in relation to the spectrum, *Toy* and *Harrison*[339] assumed that the internal photoelectric effect in silver halides was connected with the primary process of the production of the latent image. Nowadays this relationship is taken as having been proved and is regarded as evidence of the participation of photoelectrons in the process of construction of the latent image nuclei.[339-344] Measurements of the *Hall* effect,[345,346] investigations of the orientated separation of silver effected by photoelectrons in the anode region,[338,347] the bleaching-out of silver layer-wedges by positive holes in the cathode region on giving a short exposure and a synchronous direct current impulse,[348-351] the observation of the migration of a boundary layer in a galvanic cell,[352,353,353a] and other procedures[354,355] all give some idea of the properties of the electronic charge carriers. According to this the drift mobility of the photoelectrons is of the order of magnitude of 30–60 cm^2/V . sec and that of the positive holes can reach values of up to 1·1 cm^2/V . sec; the lifetime of the charge carriers is of the order of magnitude of 10^{-6} to 10^{-7} sec. Cf. also[332,356-361] as well as investigations of the influence of the crystal surface.[333,362,363] Measurements of the dark conductivity of silver halide crystals—e.g. of its relation to temperature and the influence on it of foreign ions[349,364–368]—show in addition the presence of mobile silver ions (interstitial silver ions Ag_0^+) at room temperature,[369] and these ions likewise play an important part in the primary process.

According to the Gurney-Mott theory, the process of formation of the latent image consists in the intermittent capture of the photoelectrons which are formed at the electron traps when light is absorbed[370,371] (electronic process) and the resulting attraction of the interstitial silver ions owing to the negative charging up of these traps (ionic process). The silver atoms thus formed—possibly after having been positively charged by an adsorbed silver ion, which reduces their tendency to recombine with electrons and holes[267,374,375]—also act as electron traps.[347,372,373] There is also the possibility of a further addition of photoelectrons, leading to a renewed attraction of silver ions and therefore to the formation of larger aggregates of silver. On the other hand, the positive holes, which have originated on excitation, migrate to the surface of the grain and react there with halide ions to form halogen atoms, which then pass into the gelatin.[376]

It is worth noting that the said process of latent image formation is reflected by measurements of the photoconductivity of silver halide layers. Firstly, not only the normal photoelectric effect, but a decrease in current on exposure is also observed in silver halides.[342,377] This negative photoeffect is possibly attributable to a decrease in the number of interstitial silver ions which proceeds in parallel with the formation of the latent image. Secondly, *Eggert* and *Amsler*,[344,378,379] by making electrical measurements of the photolysis of silver bromide in silver bromide photocells were able to establish that this reaction proceeded in accordance with the mechanism discussed.

Here we shall not enter into an exhaustive discussion of the entire process of latent image formation, in which a distinction has to be drawn between the initial reaction leading to a non-developable silver aggregate and the true secondary reaction in which the development nucleus is formed.[380,381] Instead, the reader is referred, for example, to the review articles by *Berg*,[331] *Hamilton* and *Urbach*,[382] which articles include discussions of the Mitchell mechanism[267,383] of latent image formation.

In spectral sensitization it is essential even in this process for the formation of the latent image to be initiated by a photoelectron which has been brought into the conduction band of the silver halide. The further process of formation might proceed in accordance with that of unsensitized emulsion layers.

The main problem to be solved as regards the mechanism of sensitization therefore lies in answering the question as to how a photoelectron can get into the conduction band of the silver halide on excitation of the dye. Amongst other things investigations of the conductivity of sensitized photoconductors are of great importance in answering this question.

4.1.2 CHARACTERISTIC FEATURES OF THE SENSITIZED PHOTOCONDUCTIVITY OF SILVER HALIDES.

4.1.2.1 *Its Relation to Wavelength.* The spectral sensitization of the photographic process and the sensitization of the photoconductivity of silver halides are closely related. This is proved by the displacement of the absorption spectrum of the photoconductivity of silver halide observed on adding sensitizing dyes. Dyeing in the spectral region of the normal absorption of the dye renders a silver halide photoconducting, i.e. beyond the absorption by the fundamental lattice in the silver halide, so that the regions of absorption, spectral photographic sensitization and photoelectric conductivity brought about by dye-sensitization coincide.

Here we shall refer to the work of *West* and *Carroll, Eggert, Terenin, Akimov et al.*, in which this parallelism between dye-absorption, photographic sensitization and sensitized photoconductivity has been demonstrated directly in the case of photographic gelatin emulsions,[36,40,296,298,384] crystalline AgBr-layers,[385-388] sublimated AgI,[389] AgBr-photoelements,[343,344,378] in Becquerel's apparatus,[390] by observations of the photoeffect in crystals,[391-396] and by means of a contactless microwave method.[397] These experiments represent a practical photoelectric analogue of the photographic sensitization reaction. Even if they do not enable any positive statements to be made regarding the individual stages themselves, that is to say, even if they are unable to decide whether the photoelectrons are formed primarily in the dye or only secondarily in the silver halide, they nevertheless give an important experimental insight into the sensitization reaction. These photoelectric investigations enable the reactions which lead to a latent image in a sensitized emulsion to be tested in a manner which is independent of the photographic process. Since the photoconductivity of other photo-semiconductors can be spectrally sensitized by dyes, and the mechanism of the spectral sensitization reaction observed in different semiconductors is undoubtedly identical, a great deal of experimental material is available for the interpretation of spectral sensitization. Most rules governing the photographic sensitization reaction also hold for the spectrally sensitized photoconductivity of other photo-semiconductors.

4.1.2.2 *The Influence of Desensitizing Dyes.* In discussing desensitization—cf. moreover Chap. III, 2.1 and[398-401]—it has already been mentioned that the investigations which have been carried out on the photoconductivity of silver halides have not yet provided conclusive evidence of an effect analogous to photographic sensitization. Whereas a desensitizing dye adsorbed to silver halide practically destroys its photographic sensitivity, according to *West* and *Carroll et al.*[36] the photoconductivity of desensitized silver halide in the spectral region of the normal absorption region of AgBr is not, or is only slightly decreased,[401] but in the region in which the dye absorbs it is increased even in such circumstances as when a sensitization reaction is taking place.[36,390,396] For instance, the crystal photoeffect in AgI is sensitized not only by Pinacyanol and other sensitizers, but also by typical desensitizers such as Phenosafranine or Methylene blue. This variable behaviour might be explained by attributing the photoconductivity of silver halide to the formation of electrons and holes. Whereas the increase in photoconductivity does not depend entirely on whether electrons or holes are present as the majority carriers, the elementary photographic process is entirely dependent on the presence of photoelectrons in the conduction band of the silver halide. Therefore any increase in the concentration of the holes in the silver halide effected by the desensitizer on irradiation in the absorption bands of the latter, can by all means lead to the sensitized photoconductivity of the silver halide, but not to the formation of a latent image. The fact that AgBr- and AgI layers are both *n*- and *p*-conducting is in agreement with this hypothesis.[402,403]

In this connection it should be pointed out that just as many electrons as positive holes (defect-electrons) are formed in a true semiconductor by means of thermal or optical excitation. It is the movement of these charge carriers under the influence of an electric field which produces conductivity. If any lattice defect centres occur in the solid, then the charge carriers obtained thermally or optically from these centres may also be made available for conduction. In the case of donor centres the conductivity is moreover carried mainly by electrons (*n*-conduction). On the other hand, in the case of acceptor centres which form mobile holes by taking up electrons (positive) holes are present as the majority carriers of the conductivity (*p*-conduction).

When AgI is sensitized by desensitizing dyes the sign of the charge carriers in the region of the absorption bands of the dye corresponds with that measured in the spectral region of the normal absorption of AgI—it is positive—whereas when it is sensitized by photographic sensitizers, a change in the sign of the majority carriers is observed: in this case AgI behaves as an *n*-conductor on exposure, thus indicating the formation of excess photoelectrons in the silver halide conduction band, this process being necessary for the production of a latent photographic image.

The observation that some sensitizers do not alter the *p*-conduction type of AgI on sensitization must not be regarded as being contradictory to these results, since the existence of *p*-conduction by no means excludes the presence of electrons in the conduction band.

From these experiments it follows therefore that:

1. Dyes (desensitizers and sensitizers) enable the photoconductivity of *p*-conducting silver halides to be sensitized without changing the type of conduction.
2. Photographic sensitizers permit a *p*-conducting silver halide to be converted into an *n*-conducting halide on irradiation in the absorption bands of the dye—corresponding to the formation of a high concentration of photoelectrons.[148]
3. Spectral photographic sensitization differs from the primary process of

desensitization in that the former reaction results in the formation of photoelectrons in the conduction band of the silver halide, whereas the latter reaction only produces an increase in the concentration of the positive holes in the silver halide.

4.1.2.3 *Supersensitization*. In contrast to desensitization, the effect of supersensitization can also be observed by making photoelectric measurements of sensitized silver halide emulsion layers. That is to say, the photoconductivity of a sensitized silver halide emulsion is greatly increased in an analogous manner to the sensitivity of the photographic emulsion[36] by the addition of a supersensitizer. Since with supersensitization of the 2nd kind, the supersensitizer has but little influence on the absorption of light by the emulsion, this signifies that supersensitization promotes an interchange between the dye and the silver halide to an equal extent in both reactions (photoconductivity and photographic effect). In addition, the fact that the supersensitizer is unable to influence the photoconductivity of the silver halide proves that the supersensitizer acts via a reaction with the dye layer and not directly on the silver halide.

This reaction between the dye and the supersensitizer will be discussed in greater detail in Chap. VI.

4.1.2.4 *Other Cases of Agreement with Photographic Effects*. The relation between photographic sensitization and the sensitized photoconductivity of a silver halide is also revealed in the following effects:

1. Not only do sensitizing dyes sensitize surface conductivity, but above all they also sensitize the photoconductivity throughout the entire volume of the AgBr microcrystals and the photographic emulsion grains.[397,404] This concurs with the findings of *Berg, Marriage* and *Stevens*,[316,405] who, by using a photographic method, were the first to demonstrate that dyes sensitized the formation of external and internal latent images in silver halide grains.

2. The photoconductivity which has been sensitized by dyes in silver halide emulsions can still be identified at −196°C, and it behaves in the same way as the sensitized photographic sensitivity in that the efficiency of the sensitizing reaction decreases by 10–20% at low temperatures.[148,406]

3. Dyes which do not spectrally sensitize the elementary photographic

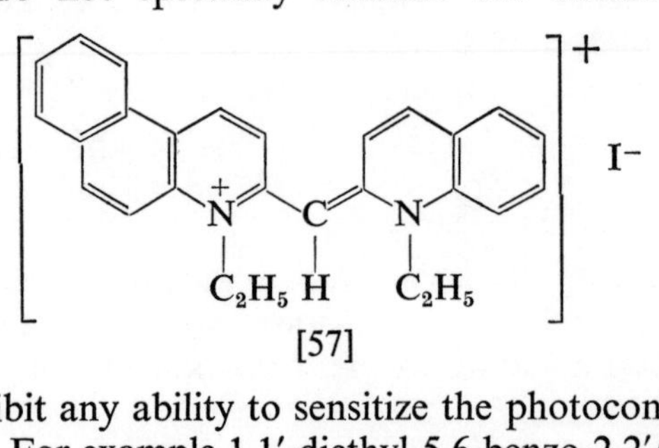

[57]

process do not exhibit any ability to sensitize the photoconductivity of photographic emulsions. For example 1,1′-diethyl-5,6-benzo-2,2′/cyanine iodide [57] in the region of 500 to 600 mμ in the spectrum represents a sensitizer both for the photographic process and for the photoconductivity of an AgBr emulsion. The dye 1,1'-diethyl-3,4-benzo-2,2'/cyanine iodide [58] which is of similar structure, does not however exhibit any sensitizing power, this being apparently due to the destruction of the co-planarity of the molecules caused by rotation of the ring systems.[36]

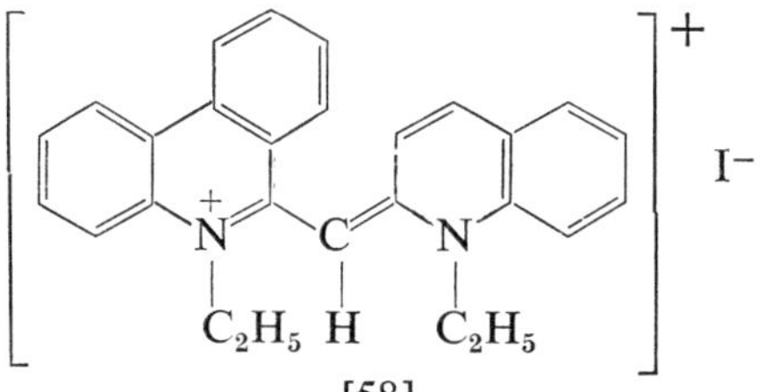

[58]

This signifies that the conditions obtaining for spectral photographic sensitization (co-planarity of the dye, close contact between it and the silver halide, etc.) must also be fulfilled for the spectral sensitization of photoconductivity to take place.

In summarizing it can be stated that the conclusion to be drawn from measurements of the photoconductivity of optically sensitized silver halide emulsion layers is that the absorption of light by the dye causes photoelectrons to be brought into the conduction band of the silver halide, and these photoelectrons lead to the formation of the latent image just as if they had been formed by the normal absorption of the silver halide. Of particular importance in attempting to clarify the mechanism of this reaction is the fact that not only can the photoconductivity of silver halide (and of photographic emulsions, respectively) be sensitized, but the photoconductivity of other photoconductors can also be sensitized by means of organic dyes.

4.2 The Spectral Sensitization of the Photoconductivity of Inorganic Photosemiconductors

4.2.1 SENSITIZABLE PHOTOSEMICONDUCTORS. Table 7 gives a compilation of the various semiconductors in which, up to the present, it has been possible to observe a spectral sensitization effect in their photoconductivity. It should be noted that not only is the sensitizability of photosemiconductors of theoretical interest, but it is also of practical importance, particularly in electrophotography. This has already been discussed in detail elsewhere.[53]

The following points should be noted regarding the methods of measurement which are given in Table 7 and which were used for investigating the photo-electric conductivity of sensitizing dyes:

Measurement of the Photo-resistance: The decrease in the resistance set up on exposing a semiconductor is generally estimated from the change in the current which occurs at a specific external voltage. Taking the measurements made on the semiconductor layers between the electrodes into consideration, the increase in conductivity can be calculated directly from readings of the photocurrent taken on a galvanometer or an electrometer. Surface type

TABLE 7

SUMMARY OF SENSITIZABLE PHOTOSEMICONDUCTORS

Substratum	Nature of the photoeffect*	Literature reference
TlCl	K	394, 407, 408
TlBr	K	251, 387, 393, 407, 409
TlI	K	388, 394, 396, 402, 403, 407–411
	I	410
NaCl	I	412, 413
HgI_2	K	387, 393, 409
CdS	K	414
	I	215, 216, 415, 416
ZnO	K	388, 396, 402, 404, 407, 408, 411, 417–419
	I	197, 394, 420–426
	E	38, 40, 42, 427–434
PbO	K	387, 393, 394, 409
Cu_2O	B	435, 436
	K	437
HgO	K	394, 438
SnO_2	K	438
CdO	K	407
Ge	K	438
AgCl	K	387, 408
AgBr	B	390
	K	251, 387, 391–393, 396, 409
	I	385, 386, 388
	P	344, 364
AgI	K	387, 393, 394, 402, 409
	I	389
	KP	251
Selenium	E,I	638a
Sulphur	E	438a

* B: Becquerel effect; K: condenser effect or Dember effect; P: photovoltaic effect; KP: change in the contact potential; I: photoconductivity in a surface type cell or in a sandwich type cell; E: electrophotographic method of measurement.

and sandwich type cells are used for the measurements (see Fig. 20). When surface type cells are used, the photoelectrically active layer is placed between two electrodes which are either mounted on glass or quartz, or are deposited directly by vaporization. When sandwich type cells are used the layer lies on a metal support and is covered with a semi-transparent evaporated electrode.

Condenser Method (Dember Effect): In this method the semiconductor in the form of a powder is placed between the two partially transparent and insulated plates of a condenser which are charged up by the charge carriers which are liberated during intermittent exposure. Since the charge carriers follow the falls in concentration produced by the absorption of light, the nature of the majority of the carriers can be deduced from the sign of the photovoltage.

Becquerel Method: In the Becquerel apparatus measurements are made of

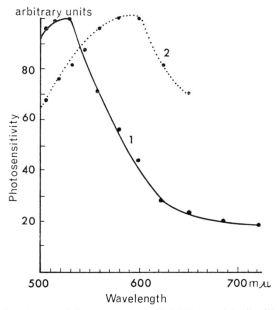

Fig. 18. The spectral distribution of photosensitivity. (1) CdS (unsensitized). (2) CdS + Merocyanine A10. (Formula, see Table 8.)

the change in the natural potential of the semiconductor immersed in an electrolyte, on exposure.

Measurement of the Photocontact Potential: In this method measurements are made of the change in the contact potential which is set up between a metal (palladium) and the semiconductor, on exposure, using for instance the Thomson vibration method. From the sign of the changes in the contact potential, conclusions can be drawn simultaneously regarding the nature of the majority of the carriers formed on exposure.[459]

Photovoltaic Effects: These effects are characterized by the development of photovoltages and photocurrents without auxiliary voltages.

Electrophotographic Methods of Measurement: In the electrophotographic process the properties of the photoconductor are estimated by measuring the fall in charge, which is induced by light, of a layer which has been charged up electrostatically by means of a corona discharge.

It can be seen from the table that the most diverse photoelectric effects in photoconducting halides, oxides, sulphides and other semiconductors can be optically sensitized by means of dyes. Figure 18 shows the effect of sensitization as exemplified by the activation spectra of an unsensitized CdS surface type cell, and of one which has been dyed by a merocyanine dye. The CdS layer of about 0·1–1 μ in thickness was obtained by evaporating highly purified CdS powder without special additions. The sensitizer was added either in solution or by sublimation in a high vacuum.[215]

It must be emphasized that the sign of the charge carriers in sensitized photoelectric effects in most cases is the same as that of the photoeffect in unsensitized

TABLE 8

SENSITIZING DYES FOR PHOTOELECTRIC EFFECTS

Dye	Charge	Type	Photographic behaviour*	Substratum	Literature
Kryptocyanine	cation	p	S	AgBr, CdS	216, 386
Fuchsine	cation	n	D	CdS	414
Malachite green	cation	n	D	ZnO, AgBr AgI	40, 46, 402 443
Erythrosine	anion	p	S	ZnO	402, 418
Eosine	anion	p	S	ZnO	403, 420
Fluorescein	anion	p	S	ZnO	427
Rhodamine B	cation	n	S/D	TlI, CdS	215, 402
Phthalocyanine	non-ionic	p	S	ZnO, HgO	418, 444
Chlorophyll	,,	p	S	ZnO, HgO	417, 438, 444
Merocyanines†	,,	p	S	CdS	215

* S: Sensitizer; D: Desensitizer.   † Formulae [59–62].

semiconductors. This rule was first established by *Terenin, et al.*[251,394,402,439–441] by means of the Dember effect (by determining the polarity of the diffusion photo-EMF) and by observing the changes in the contact potential of sensitized photosemiconductors on exposure.

For example the majority carriers are electrons in sensitized ZnO layers (n-type) as well as in the spectral region of the absorption of the dye and again in the region of the normal absorption of ZnO.[402,408] In thallium halides, which are included in the p-conductors, the sign of the majority carries in the region of the normal absorption and that of the absorption of the sensitizer corresponds to hole conduction, and is therefore positive.

In this connection we would however refer to an observation which is important in regard to photographic sensitization, namely that *p*-conducting silver halide layers and crystals become *n*-conducting on sensitization and exposure in the absorption region of the dye. This was not only established by means of measurements of the crystal photoeffect,[402] but *West, et al.*[148,422] also made direct observations of the migration of photoelectrons in highly purified, dyed, silver bromide crystals using the impulse field method.[374]

4.2.2 THE NATURE OF SENSITIZING DYES. In order to demonstrate the large variety of dyes which can be used for the spectral sensitization of photoconductivity, a few typical sensitizing dyes have been grouped together in Table 8.

Formulae of the merocyanines in Table 8:

Merocyanine A 10 [59]

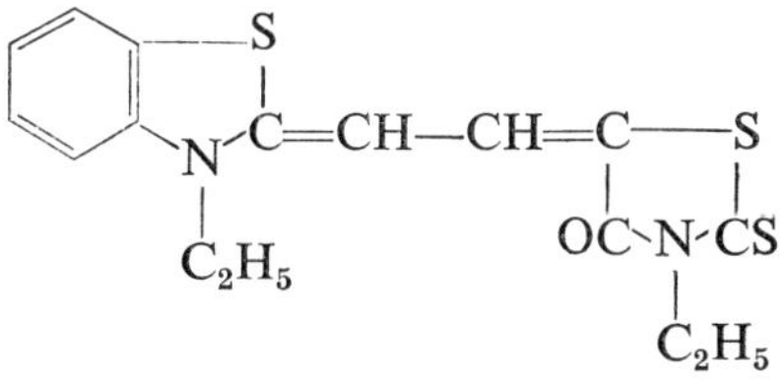

Merocyanine FX 79 [60]

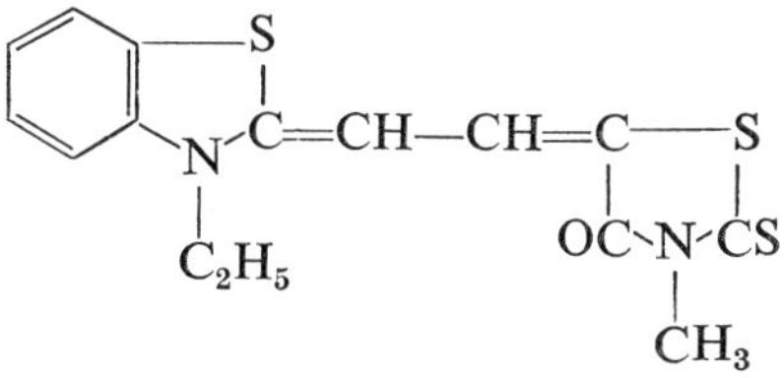

Merocyanine FX 4 [61]

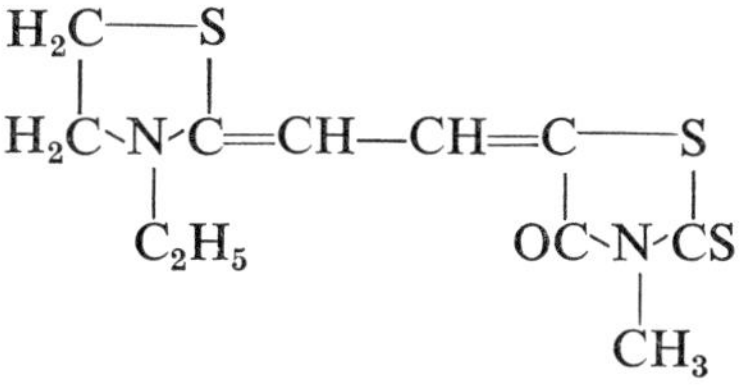

Merocyanine FX 30 [62]

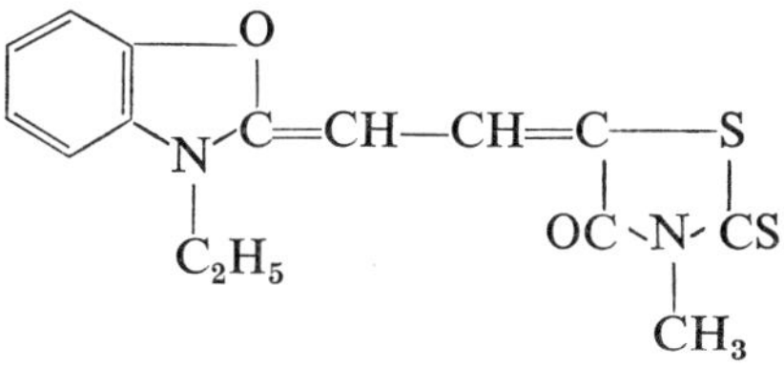

This compilation which covers only a few of the dyes studied reveals the following information:

1. Anionic, cationic and nonionic dyes can sensitize the photoconductivity of various semiconductors.

2. Not only photographic sensitizers but also desensitizers have a sensitizing action on the photoconductivity of photoconductors. For example the photoconductivity of AgI and TlI can be sensitized by photographic sensitizers such as Eosine, Erythrosine and Pinacyanol and by desensitizers such as Phenosafranine, Malachite green, etc.[36,402] The cause of this departure from the photographic behaviour may be attributed to the fact that the formation of a latent image presupposes the excitation of electrons in the conduction band of the silver halide, whereas the photoconductivity of *p*-conducting halides may be due to the formation of holes.
3. Dyes of the *n*- and *p*-conduction types can be used for spectral sensitization. However no direct connection appears to exist between the type of conduction in a dye and that in a sensitized photosemiconductor. For instance the photoconduction in *n*-conducting CdS, as is shown by the activation spectra in Fig. 19, can be sensitized by Rhodamine B (*n*-type) and by merocyanine dyes (*p*-type).

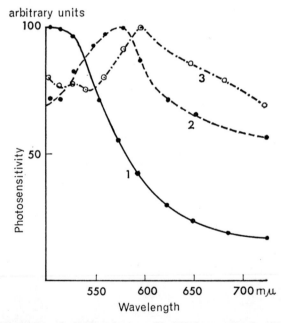

Fig. 19. Spectral distribution of photosensitivity. (1) CdS (unsensitized). (2) CdS + Merocyanine FX 79 (*p*-type; formula, see Table 8). (3) CdS + Rhodamine B (*n*-type).

4.2.3 LAWS GOVERNING THE SPECTRAL BEHAVIOUR OF PHOTOSEMICONDUCTORS. The spectral sensitization of the photoconductivity of various inorganic photosemiconductors can be seen to conform with a number of laws which are also characteristic of photographic sensitization. This conformity forms the basis for ultimately concluding that photographic and photoelectric sensitization reactions are identical, such a conclusion being of decisive importance as regards the theory of the mechanism of sensitization: for the primary process of energy transfer might at times take place in a similar manner. In the case of silver halides—and possibly of other photosensitive systems as well—this reaction is followed immediately by a chemical process which leads to stable photolytic changes.

4.2.3.1 *Sensitization Yield.* The photoconductivity of sensitized semiconductors which can be measured in the spectral region in which the dye absorbs can reach up to 100% of that of an unsensitized photoconductor (CdS, TlI, ZnO).[208,402,408,415] Since the photoconductivity is related to the same number of absorbed photons, this signifies that in favourable cases the energy absorbed by the dye will be used to yield the same number of electronic charge carriers as when it is absorbed directly by the photoconductor.

In sensitizing with infra-red sensitizers however—e.g. in the system CdS/Kryptocyanine—a transfer yield of only 5% could be measured. It is however worth noting that the sensitization yield in infra-red sensitized photographic emulsion layers is also only small.

4.2.3.2 *Repeated Sensitizing Power.* As in spectral photographic sensitization, 10^2 to 10^3 photoelectrons are also formed from one dye molecule in the case of sensitized semiconductors. *Matejec*[422,445] observed, for instance, that ZnO which had been spectrally sensitized by Erythrosine gave a yield of up to 250 photoelectrons per dye molecule.

In this connection it should be pointed out that the repeated reactivity of dye molecules which is proved by these high yields—measurements of which it is true have given results of a similar order of magnitude in the case of photographic emulsions[209,211,221,446]—is not necessarily an absolute prerequisite for the production of the latent image in practical photography. For, whereas about 10^5 to 10^6 dye molecules are adsorbed to the silver halide grains which are about 1 μ^2 in size, the absorption of 4 to 50 quanta—i.e. dye molecules—per grain, according to the sensitivity of the emulsion, is already sufficient[319,328] to render a grain developable.

For the photolysis of silver bromide, in which a fairly large quantity of silver is separated out (print-out effect), a repeated transfer of photons by means of the adsorbed dye molecules, i.e. regeneration of the dye molecules after the transfer of energy is however necessary.

4.2.3.3 *The Effect of Temperature.* The influence of temperature, as indicated by the decrease in the photographic sensitivity of a sensitized silver halide emulsion on lowering the temperature is also observed in the case of sensitized photosemiconductors. For example the photoeffect in a layer of TlI which has been sensitized by Methylene blue in the region of absorption of the dye increases tenfold in passing from $-50°$ to $+70°C$, whereas the effect on exposure in the region in the spectrum of the normal absorption of TlI increases by only a few percent.[402] A few results which are contradictory to this finding[404,410] suggest however that this effect should be further investigated.

4.2.3.4 *The Adsorption of Dyes.* The adsorption of various dyes not only to silver halides but also to photosemiconductors will be described by means of the Langmuir isotherm. According to this equation Erythrosine and Fluorescein are adsorbed from alcoholic solution to ZnO, so that it may be concluded that the zinc oxide surface is covered with a monomolecular layer of dye.[419] The maximum photoeffect is obtained when 30–40% of the ZnO surface is covered with a monomolecular layer of dye. In the case of AgBr grains the best photographic sensitizing effect is observed when the coverage exceeds 50%.[230]

As with the silver halides, various dyes can be seen to show departures from the *Langmuir* adsorption isotherms. Trypaflavine for instance follows *Freundlich*'s equation corresponding to an adsorption in multimolecular layers. The measurements made by *Grossweiner, et al.*[420,415] however, show that even these multimolecular dye layers sensitize. We were still able to achieve sensitization of the photoconductivity of CdS with layers of dye of $0.1-1\,\mu$ in thickness (layers of several hundreds of molecules) (cf. Fig. 18).

As with photographic sensitization, the *H*- and *J*-maxima which are characteristic of aggregated dyes, also occur in the activation spectra of the sensitized photoconductivity produced by certain dyes (cyanines, etc.).[391,393,407,408,417,425]

4.2.3.5 *Desensitization.* Up to the present only *Markevitsch* and *Putzeiko*[419] have reported on the desensitization of the photoconductivity of a semiconductor. The authors observed a decrease in the photoconductivity and in the photo-EMF (Dember effect) in sensitized ZnO after the optimum dye concentration had been exceeded. It still remains uncertain however whether we may not be concerned here with a filtering action of the dye, since the observations mentioned in the last paragraph did not reveal any desensitizing effect. It must also be remembered that the sensitized photoeffect, i.e. the efficiency of the transfer of energy in the region of sensitization, generally decreases with increase in the thickness of the adsorbed layer of dye. This might be considered to be not so much a case of desensitization as of the sensitization behaving in conformity with a certain law.

4.2.3.6 *The Influence of Gases.* The efficiency of the spectral sensitization of photoelectric effects in semiconductors—$AgBr$,[251,447] AgI,[392–394,402,448] $TlBr$,[251,447] TlI,[392–394,402,411,448] CuI,[394] ZnO[394,402,411,421,449,450]—is in some cases raised considerably by the addition of gases and vapours of molecules possessing electron affinity, such as O_2, Br_2, I_2, *p*-benzoquinone, etc. For instance bromine is observed to have a strong optical sensitizing action on the crystal photoeffect in polycrystalline layers of TlBr which have been dyed by Malachite green, Crystal violet or thionine. This optical sensitization regresses when the bromine is desorbed.[251]

In contrast to compounds with electron affinity, H_2, NH_3 or H_2O often remain without any perceptible influence on sensitized photoconductivity;[425] cf. however also *Akimov, et al.*[411,421]

4.2.3.7 *Supersensitization.* As in the case of photographic sensitivity, the photoconductivity of spectrally sensitized semiconductors can also be greatly increased in the spectral region of the dye by the addition of a supersensitizer.

The effect has been studied above all with silver halides—see Chap. III, 1.2.3—and with ZnO. Two of the results obtained with ZnO are of particular interest.

1. As with the supersensitizing effect in photographic emulsions, a large increase in the sensitized photoconductivity can be achieved by mixing two dyes—e.g. by a combination of 1,1'-diethyl-2,2'-quinocyanine iodide and 3-ethyl-2-(*p*-dimethylaminostyryl)benzothiazole iodide.[425]

2. The photoconductivity of ZnO in the region of its normal absorption in the spectrum is greatly increased by interaction between the electron acceptors (*p*-chloranil, phthalic anhydride, etc.) and the sensitizing dyes (Rose Bengals, etc.) adsorbed to the microcrystalline ZnO.[69,424,451,452] It is therefore natural to link this supersensitizing effect with the changes in the sensitized photoconductivity observed during the action of gases with electron affinity. The main point to be emphasized is that a supersensitizer does not become active until a dye is present, in other words, it must react with the dye molecules in some way which will be favourable to the transfer of energy. In a later section, when dealing with the effects of doping the dyes, it will be shown that a great deal is already known about this reaction.

4.2.3.8 *The Quenching of Fluorescence.* It has already been pointed out in Chap. III, 1.2.8 that a decrease in the fluorescence yield of sensitizers is connected with the sensitized photolysis of silver halides. The same effect is also observed when the dye is adsorbed to a photosemiconductor which can be spectrally sensitized—AgBr, AgI, TlBr, TlI, ZnO, PbO.[251,253,254,453] Quenching of the fluorescence does not however occur on adsorption of the dye to compounds such as MgO, BaO, SiO_2, TiO_2, etc. which are not altered photochemically or which do not exhibit any photoelectric effects.

From these observations it can be deduced that any possible return with emission of fluorescence of an excited electron into the adsorbed dye does not occur if photochemical decomposition or a photoelectric effect is produced in the substratum as a result of the transfer of electrons or because of a direct transfer of energy. Since the fluorescence radiation is emitted within 10^{-9} seconds, it also follows from the quenching of the fluorescence that the sensitization reaction takes place in less than 10^{-9} seconds.

However, recent studies by *Gilman*[453a] on the low-temperature emission spectra of sensitising dyes (1,1'-diethyl-2,2'-cyanine *et al.*) adsorbed to silver halide in gelatin show that there is the possibility that electrons—which are transferred from the dye into the conduction band of the silver halide—may return from the silver halide to the dye. *Gilman* observed that the transition of the *adsorbed* dye from the monomeric to the *J*-aggregate state is connected with a loss of the monomeric phosphorescence and the formation of a new phosphorescent emission at a longer wavelength. This new delayed luminescence is in good experimental agreement with a mechanism according to which electrons from the dye are temporarily freed in the conduction band of the substratum and recombine with dye positive holes trapped in the *J*-aggregate. *Gilman* was able to prove that the second-order recombination process led to the formation of excited *J*-aggregate triplets which radiatively decay. There seems to be no doubt that this electron hole recombination which is connected with the *J*-aggregate state and which can be observed directly by low-temperature luminescence may also be effective during the spectral sensitization process.

4.2.4 PHOTOINDUCED EPR SIGNALS. $AgBr$[454-456] and ZnO[457] which have been sensitized by dyes give sharp paramagnetic resonance absorption lines on exposure in the region in the spectrum in which the respective dyes absorb; AgBr for instance gives a singlet EPR line.

These EPR signals may either be taken to indicate that dye radicals are formed, possibly by the transfer of electrons during the sensitization reaction, or else they may be attributed to the bromine or oxygen molecules which are liberated during the photolytic reaction and the process of desorption, respectively.

Terenin, et al.[404,456] regarded the second possibility as being the more probable for AgBr, since the EPR line can also be observed when AgBr is exposed in its normal absorption region, and again with dyes on the addition of bromine. However further investigations of EPR appear to be necessary in order to establish this assumption on a firm basis.

In this connection the measurements of the electron spin resonance spectra of polycrystalline donor-acceptor complexes between sensitizing dyes and acceptor compounds—made by Lu Valle and co-workers[457a]—should be mentioned. Since it is possible that the dye may form eventually a donor-acceptor complex with lattice silver ions, impurities or any components there should be a correlation between the intensity of the EPR spectrum of the complex and the ability of the donor to sensitize. It is worth noting that the spectra seem to give some indication of such a correlation.

4.3 *The Spectral Sensitization of the Photoconductivity of Organic Semiconductors*

Since the conductivity of organic solids is generally due to the formation and migration of electronic charge carriers—cf. for instance[53,458-460]—the photoconductivity of these photosemiconductors can likewise be spectrally sensitized by means of organic sensitizing dyes.

This effect firstly enables the experimental material on the spectral sensitization reaction to be extended and many of the laws governing the behaviour of inorganic semiconducting substances to be supplemented and confirmed. Secondly the effect makes it possible first and foremost to extend the use of organic photosemiconductors which are admirably suitable for electrophotographic image recording, but which frequently only absorb in the UV region, to the visible region of the spectrum.[55,461-463]

Special attention is drawn to the following facts:

1. The various classes of organic photosemiconductors (hydrocarbons, charge-transfer complexes and high polymers) can be spectrally sensitized by the addition of a dye. Table 9 contains several examples.

TABLE 9

Organic semiconductors	Type	Sensitizing dye	Literature
Anthracene	p	Rhodamine B	464–466
Cu-phenylacetylide PAC [63]	p	Pinacyanol, Methylene blue	404, 467, 468
Poly-acetylene	p	Mg-Phthalocyanine	404, 467
Gelatin		Eosine	469
Sulphydryl-protein-complex			470
Poly-N-vinyl-carbazole [64]	p	Rose Bengals	43, 471, 472
Poly-N-vinyl-carbazole + doping agent	p, n	Methyl violet	473
Thiadiazoles			474
Organic electrophotographic material [65]		Rhodamine B	475
Proteins		Fluorescein	476

Formulae of the compounds [63] [64] and [65] in Table 9:

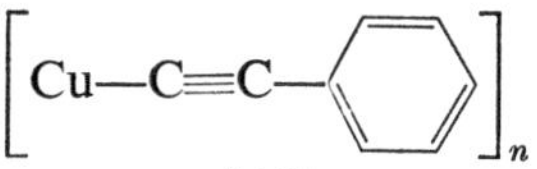

$$\left[Cu-C\equiv C-\phenyl \right]_n$$

[63]

$$\left[\text{carbazole}-CH-CH_2- \right]_n$$

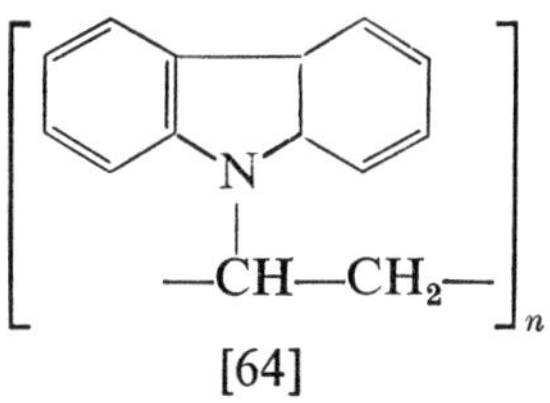

[64]

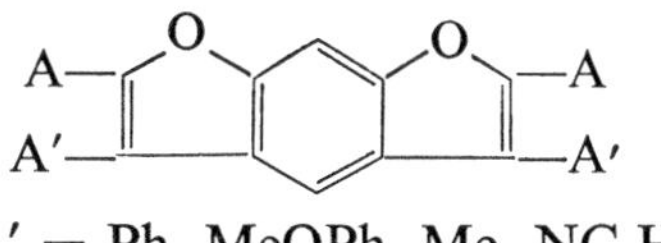

AA′ = Ph, MeOPh, Me$_2$NC$_6$H$_4$

[65]

2. The same dyes can be employed for sensitizing organic photosemiconductors as those which are used for inorganic semiconductors. That is to say, anionic, cationic and neutral dyes (Erythrosine, Eosine, Rose Bengals; Pinacyanol, Methylene blue, Rhodamine B; chlorophyll, Mg-Phthalocyanine, etc.) of the n- and p-conduction types can be used.

3. The sensitizing power of a dye is greatly dependent on its adsorption.[467] Moreover dyes sensitize both when they are deposited in the form of aggregates as well as when they are adsorbed in a monomolecular layer.

4. Up to the present, studies of the sensitization effect have been made chiefly with p-conducting organic photosemiconductors. In other words, the sensitization reaction of these photoconductors cannot be attributed to the transfer of electrons from the dye to the substratum, but in analogy to the p-conducting thallium halide, to an increase in the concentration of positive holes in the photoconductor.

In the case of n-conducting sensitized organic photoconductors the concentration of electrons in the semiconductor is however raised by the sensitization reaction.

In summarizing it may be said that studies of organic photosemiconductors likewise indicate that the concentration of positive holes or of electrons—depending on the type of conduction—in a photosemiconductor is raised by dye sensitization on irradiation in the region of the spectrum in which the dye absorbs.

References

[309] GURNEY, R. W. and MOTT, N. F., *Proc. roy. Soc.*, Ser. A, **164**, 151 (1938).

[310] MOTT, N. F., *Photogr. J.*, B. **88**, 119 (1948).

[311] MOTT, N. F., *Photogr. J.*, **81**, 62 (1941); *J. Phys. Radium*, (2), **7**, 249 (1946).

[312] MOTT, N. F. and GURNEY, R. W., *Electronic Processes in Ionic Crystals*, Clarendon Press (Oxford) 1940.

[313] EGGERT, J. and NODDACK, W., *Z. Phys.*, **58**, 861 (1929).

[314] MEIDINGER, W., *Z. wiss. Photogr.*, **44**, 1, 137 (1949).

[315] TREICHEL, U., *Veröffentl. wiss. Photolab. Agfa Wolfen.*, **9**, 51 (1961).

[316] BERG, W. F., MARRIAGE, A. and STEVENS, G. W. W., *J. opt. Soc. Amer.*, **31**, 385 (1941).

[317a] NODDACK, W., SCHALLER, H. and HECKER, W., *Wissenschaftliche Photographie* (*Scientific Photography*) (Int. Konf. Köln., 1956) edited by O. HELWICH, p. 82 (Darmstadt) 1958.

[317b] SWINNERTON, A. J., *J. photogr. Sci.*, **10**, 217 (1962).

[317c] HILLSON, P. J. and SUTHERNS, E. A., *The Theory of the Photographic Process*, 3rd edn. edited by C. E. K. MEES and T. H. JAMES, p. 120, Macmillan (New York, London) 1966.

[318] FARNELL, G. C. and CHANTER, J., *J. photogr. Sci.*, **9**, 73 (1961).

[319] MARRIAGE, A., *J. photogr. Sci.*, **9**, 92 (1961).

[320] BUSSE, J. and HENNIG, K., *Phys. stat. Sol.*, **7**, K 83 (1964).

[321] GREENSLADE, D. J., *Brit. J. appl. Phys.*, **16**, 1921 (1965).

[322] CALVIN, S. G., *U.S. At. Energy Comm. NYO*, 2471-5 (1965).

[323] BELOUS, V. M. and CHIBISOV, K. V., *Zh. nauch. prikl. Fotogr. Kinem.*, **8**, 334 (1963).

[324] BELOUS, V. M. and CHIBISOV, K. V., *Dokl. Akad. Nauk SSSR*, **156**, 121 (1964).

[325] BELOUS, V. M., BUGRIENKO, V. I. and GOLUB, S. I., *Optika Spektrosk.*, **17**, 406 (1964).

[326] KLEIN, E., *Z. Elektrochem.*, **62**, 505 (1958).

[327] KLEIN, E. and MATEJEC, R., *Z. Elektrochem.*, **63**, 883 (1959); *Naturwissenschaften*, **46**, 225 (1959).

[328] WAIDELICH, W., *Z. wiss. Photogr.*, **56**, 97 (1961).
[329] WAIDELICH, W. and FUCHS, W., *Photogr. Korr.*, **98**, 30 (1962).
[330] WOLFF, H., *Fortschr. chem. Forsch.*, **2**, 375 (1952).
[331] BERG, W. F., *Photographic Science*, Symposium Zürich 1961, edited by W. F. BERG, p. 27, Focal Press (London) 1963.
[332] SIMON, H. and SUHRMANN, R., *Der lichtelektrische Effekt und seine Anwendungen* (*The Photoelectric Effect and its Applications*), Springer (Berlin) 1958.
[333] SNAVELY, B. A., *Phys. stat. sol.*, **9**, 709 (1965).
[334] BALLONOFF, A., *Phys. Rev.*, **137**, A, 462 (1965).
[335] LOSKUTOV, K. N., *Fiz. tverd. Tela*, **8**, 959 (1966).
[336] VACEK, K., *Czech. J. Phys.*, **15**, 940 (1965).
[337] CHOLLET, L. and ROSSEL, J., *Helv. phys. Acta*, **33**, 627 (1960).
[338] HAYNES, J. R. and SHOCKLEY, W., *Physiol. Rev.*, **82**, 935 (1951).
[339] TOY, F. C. and HARRISON, G. B., *Proc. roy. Soc.*, Ser. A, **127**, 613 and 629 (1930).
[340] BARSHCHEVSKII, B., *Zh. eksp. theor. Fiz.*, **16**, 815 (1946).
[341] LEHFELDT, W., *Nachr. Ges. Wiss. Göttingen, Math. Phys. Kl.*I 263 (1933).
[342] MATEJEC, R., *Z. Phys.*, **148**, 454 (1957).
[343] EGGERT, J. and AMSLER, H., *Z. Elektrochem.*, **56**, 733 (1952).
[344] EGGERT, J., *Angew. Chem.*, **73**, 417 (1961).
[345] HANSON, R. C. and BROWN, F. C., *J. appl. Phys.*, **31**, 210 (1960); *J. chem. Phys.*, **66**, 2376 (1962).
[346] MASUMI, T., AHRENKIEL, R. K. and BROWN, F. C., *Phys. stat. sol.*, **11**, 163 (1965).
[347] HAYNES, J. R. and SHOCKLEY, W., *Report on a Conference on the Strength of Solids*, Univ. Bristol, 1947. The Physical Society, London, 1948, p. 151.
[348] MALINOWSKI, J. and SÜPTITZ, P., *Z. wiss. Photogr.*, **57**, 4 (1963).
[349] HEUKEROTH, U. and SÜPTITZ, P., *Phys. stat. sol.*, **13**, 285 (1966).
[350] IRMER, J., *Phys. stat. sol.*, **2**, 1552 (1962).
[351] MALINOWSKI, J. and PLATIKANOWA, V., *Izv. Inst. Fizikokhim., Bulg. Akad. Nauk*, **5**, 97 (1965); *Phys. stat. sol.*, **6**, 885 (1964).
[352] JAENICKE, W., *Phys. stat. sol.*, **3**, 31 (1963).
[353] JAENICKE, W. and EISENMANN, E., *Naturwissenschaften*, **52**, 491 (1965).
[353a] EISENMANN, E. and JAENICKE, W., *Photogr. Sci. Eng.*, **11**, 173 (1967).
[354] LUCKEY, G. W., *Disc. Faraday Soc.*, **28**, 113 (1959).
[355] KNAPTON, J. D., *Chem. Abstr.*, **65**, 12, 15661 (1965).
[356] AHRENKIEL, R. K. and VAN HEYNINGEN, R. S., *Phys. Rev.*, **144**, (2), 576 (1966).
[357] WEBB, J. H., *J. Appl. Phys.*, **26**, 1309 (1955).
[358] SÜPTITZ, P., *Phys. stat. sol.*, **1**, K 81 (1961).
[359] SAUNDERS, V. I., TYLER, R. and WEST, W., *Photographic Science*, Symposium Zürich 1961, edited by W. F. BERG, p. 22, Focal Press (London) 1963.
[360] SÜPTITZ, P., *Z. Elektrochem.*, **63**, 400 (1959).
[361] LAYER, H., MILLER, M. G. and SLIFKIN, L., *J. appl. Phys.*, **33**, 478 (1962).
[362] BRADY, L. E. and HAMILTON, J. F., *J. appl. Phys.*, **36**, 1439 (1965).
[363] GEORGIEV, M., *Phys. stat. sol.*, **15**, 193 (1966).
[364] KOCH, E. and WAGNER, C., *Z. phys. Chem.*, **38B**, 295 (1937).
[365] TELTOW, J., *Ann. Phys.*, **6**, 563 (1950); **16**, 268 (1955).
[366] STASIW, O., *Z. Phys.*, **127**, 522 (1950).
[367] MATEJEC, R., *Photogr. Korr.*, **93**, 17 (1957); *Mitt. ForschLab. Agfa Leverkusen-München*, Vol. III, p 45, Springer (Berlin) 1961.
[368] MÜLLER, P., *Phys. stat. sol.*, **9**, K 193 (1965).
[369] HAMILTON, J. F., and BRADY, L. E., *J. appl. Phys.*, **30**, 1893 (1959).
[370] BRADY, L. E. and HAMILTON, J. F., *J. appl. Phys.*, **37**, 2268 (1966).
[371] NORIAKI, I., KENJI, K. and TOKUO, S., *Photogr. Sensitivity* (Tokyo), **3**, 15, 21 (1963).
[372] GRIMLEY, T. B. and MOTT, N. F., *Disc. Faraday Soc.*, **1**, 3 (1947).
[373] YAMADA, K. and OKA, S., *Naturwissenschaften*, **43**, 175 (1956).
[374] MITCHELL, J. W., *Rep. Progr. Phys.*, **20**, 433 (1957).
[375] MATEJEC, R., *Naturwissenschaften*, **43**, 533 (1956); *Z. Phys.*, **148**, 454 (1957).
[376] HAMILTON, J. F., BRADY, L. E. and HAMM, F. A., *J. appl. Phys.*, **29**, 800 (1958).
[377] MEIER, H., *Z. wiss. Photogr.*, **53**, 117 (1959).
[378] AMSLER, H., *Z. Elektrochem., Ber. Bunsenges. phys. Chem.*, **57**, 801 (1953).
[379] EGGERT, J., *Schweiz. Photorundschau*, **15**, 391 (1961).
[380] BURTON, P. C. and BERG, W. F., *Photogr., J.*, **86B**, 2 (1946).

[381] BERG, W. F. and BURTON, P. C., *Photogr. J.*, **88B**, 84 (1948).
[382] HAMILTON, J. F. and URBACH, F., *The Theory of the Photographic Process*, 3rd edn., edited by C. E. K. MEES and H. T. JAMES, p. 87, Macmillan (New York, London) 1966.
[383] HAMILTON, J. F. and BRADY, L. E., *Photogr. Sci. Eng.*, **8**, 189 (1964).
[384] GROSS, L. G., *Zh. nauch. prikl. Fotogr. Kinem.*, **5**, 54 (1960)
[385] KAMEYAMA, N. and MIZUTA, K., *J. Soc. chem. Ind., Japan*, **42**, 426B (1939).
[386] KAMEYAMA, N. and FUKOMOTO, T., *J. Soc. chem. Ind., Japan*, **42**, 244B (1939).
[387] AKIMOV, I. A., *Dokl. Akad. Nauk SSSR*, **137**, 624 (1961).
[388] AKIMOV, I. A. and PUTZEIKO, E., *Fiz. tverd. Tela*, **4**, 1542 (1962).
[389] GOLDMAN, A. and AKIMOV, I., *Zh. fiz. Khim.*, **27**, 355 (1953).
[390] SHEPPARD, S. E., VANSELOW, W. and HAPP, G. J., *J. phys. Chem.*, **44**, 411 (1940); **53**, 331 and 1403 (1949).
[391] PUTZEIKO, E. and TERENIN, A., *Zh. fiz. Khim.*, **23**, 676 (1949).
[392] AKIMOV, I. and PUTZEIKO, E., *Dokl. Akad. Nauk SSSR*, **102**, 481 (1955).
[393] AKIMOV, I., *Zh. fiz. Khim.*, **30**, 1007 (1956).
[394] TERENIN, A., PUTZEIKO, E. and AKIMOV, I., *J. Chim. phys.*, **54**, 716 (1957).
[395] AKIMOV, I. A. *Fiz. tverd. Tela*, **4**, 1549 (1962).
[396] AKIMOV, I. A., *Dokl. Akad. Nauk SSSR*, **164**, 533 (1965).
[397] BARANOV, E. and AKIMOV, I., *Dokl. Akad. Nauk SSSR*, **154**, 184 (1964).
[398] DÄHNE, S., *Mobr. dtsch. Akad. Wiss., Berlin*, **5**, 454 (1963).
[399] KHEIMAN, A. S., NATANSON, S. V. and DONATOVA, V. P., *Zh. nauch. prikl. Fotogr. Kinem.*, **9**, 216 (1964).
[400] TAMURA, M., HADA, H. and TANAKA, T., *Nippon Sahshin Gakkai Kaishi*, **27**, 34 (1964).
[401] DEMIDOV, K. K. and POZIGUN, E. A., *Zh. nauch. prikl. Fotogr. Kinem.*, **6**, 161 (1961).
[402] TERENIN, A., PUTZEIKO, E. and AKIMOV, I., *Wissenschaftliche Photographie (Scientific Photography)* (Int. Konf. Köln., 1956), edited by O. HELWICH, p. 117 (Darmstadt) 1958.
[403] BOURDON, J., *J. phys. Chem.*, **69**, 705 (1965).
[404] TERENIN, A. and AKIMOV, I., *J. phys. Chem.*, **69**, 730 (1965).
[405] BERG, W. F., MARRIAGE, A. and STEVENS, G. W., *Photogr. J.*, **81**, 413 (1941).
[406] SHEPPARD, S. E., WIGHTMAN, E. P. and QUIRK, R. F., *J. phys. Chem.*, **38**, 817 (1934).

[407] PUTZEIKO, E. and TERENIN, A., *Dokl. Akad. Nauk SSSR*, **90**, 1005 (1953).
[408] TERENIN, A. and AKIMOV, I. A., *Z. phys. Chem. Leipzig*, **217**, 307 (1961).
[409] AKIMOV, I., *Photogr. Sci. Eng.*, **3**, 197 (1959).
[410] AKIMOV, I., *Dokl. Akad. Nauk SSSR*, **151**, 310 (1963).
[411] AKIMOV, I. and TERENIN, A., *Dokl. Akad. Nauk SSSR*, **135**, 109 (1960).
[412] YAGI, H., *Mem. Coll. Sci. Kyoto*, **A26**, 75 (1960).
[413] RAKITYANSKAYA, O. F., *Zh. fiz. Khim.*, **38**, 1008 (1964).
[414] CHALILOV, A. CH. and ISAEV, F. K., *Dokl. Akad. Nauk Azerbaidzhan*, **2**, 75 (1960).
[415] NELSON, R. C., *J. opt. Soc. Amer.*, **51**, 1182 (1961).
[416] MEIER, H., *J. phys. Chem.*, **69**, 719 (1965).
[417] TERENIN, A. and PUTZEIKO, E., *J. Chim. phys.*, **55**, 681 (1958).
[418] PUTZEIKO, E., *Dokl. Akad. Nauk SSSR*, **91**, 1070 (1953); **129**, 303 (1959).
[419] MARKEWITSCH, N. N. and PUTZEIKO, E., *J. Chim. phys.*, **36**, 2393 (1962).
[420] DUDKOWSKI, S. and GROSSWEINER, L. I., *J. opt. Soc. Amer.*, **54**, 486 (1964).
[421] KORSUNOVSKII, G. A., *Dokl Akad. Nauk SSSR*, **134**, 1394 (1960).
[422] MATEJEC, R., *Z. Elektrochem.*, **65**, 183 (1961).
[423] HISHIKI, Y., TAMURA, H., NAMBA, S. and TAKI, K., *Rept. Inst. Phys. Chem. Res., Tokyo*, **36**, 386 (1960).
[424] NAMBA, S. and HISHIKI, Y., *Rept. Inst. Phys. Chem. Res., Tokyo*, **39**, 27 (1963).
[425] NAMBA, S. and HISHIKI, Y., *J. phys. Chem.*, **69**, 774 (1965).
[426] CULVER, R. B. and SORENSEN, D. P., *J. chem. Phys.*, **42**, 2975 (1965).
[427] GERRITSEN, H. J., RUPPEL, W. and ROSE, A., *Helv. phys. acta*, **30**, 504 (1957).
[428] ZELEVIC, I., KALINANSKIENE, B., NYUNKO, L., PLAVINA, I. and SUVEIZDIS, E., *Elektrofotogr. Magnitogr.*, 17 (1959).
[429] GRENISHIN, S. G. and CHERKASOV, A., *Zh. nauch. prikl. Fotogr. Kinem.*, **5**, 433 (1960).
[430] BICKMORE, J. T., HAYFORD, R. E. and CLARK, H., *Photogr. Sci. Eng.*, **3**, 210 (1959).
[431] U.S.P. 3,052,540, H. G. GREIG (1962).
[432] U.S.P. 3,238,149 (Cl. 252–501) SPURR, R. L., March 1 (1966).
[433] Belg. P. 666,076, I. PANKEN (to Ing. C. Olivetti + C.) Dec. 29 (1965).
[434] F.P. 1,296,136 dated 1/6.1961; issued 9/11.1962, W. GESIERICH, F. NISSL and G. SCHWARZKOPF (to Agfa-Akt. Ges., Leverkusen).

[435] MOSER, J., *Mh. Chem.*, **8**, 373 (1887).
[436] RIGOLLOT, H., *C.R. Acad. Sci., Paris*, **116**, 873 (1893).
[437] KARKHANIN, YU. I., KOSHEVIN, I. V. E. and PEKA, G. P., *Ukr. fiz. Zh.*, **5**, 809 (1960).
[438] TERENIN, A. and PUTZEIKO, E., *Angew. Chem.*, **70**, 508 (1958).
[438a] ING, S. W. and CHIANG, Y. S., *J. Chem. Phys.*, **46**, 487 (1967).
[439] PUTZEIKO, E., *Dokl. Akad. Nauk SSSR*, **67**, 1009 (1949).
[440] AKIMOV, I., *Dokl. Akad. Nauk SSSR*, **128**, 691 (1959).
[441] TERENIN, A., PUTZEIKO, E. and AKIMOV, I., *Disc. Faraday Soc.*, **27**, 83 (1959).
[442] SAUNDERS, V. I., TYLER, R. W. and WEST, W., *Photographic Science*, Symposium Turin 1963, edited by G. SEMERANO and U. MAZZUCATO, p. 73, Focal Press (London) 1965.
[443] WEIGL, J. W., *Photographic Science*, Symposium Zürich 1961, edited by W. F. BERG, p. 354, Focal Press (London) 1963.
[444] PUTZEIKO, E., *Radiotekh. Elektron.*, **1**, 1127 (1956).
[445] MATEJEC, R., *Photogr. Korr.*, **97**, 51 (1961).
[446] NATANSON, S. V., *J. Chim. phys.*, **14**, 278 (1940).
[447] AKIMOV, I. A., *Dokl. Akad. Nauk SSSR*, **121**, 311 (1958).
[448] PUTZEIKO, E., *Dokl. Akad. Nauk SSSR*, **78**, 453 (1951).
[449] PUTZEIKO, E., *Radiotekh. Elektron.*, **1**, 1364 (1956).
[450] PUTZEIKO, E. and TERENIN, A., *Dokl. Akad. Nauk SSSR*, **101**, 645 (1955).
[451] INOUE, E., YAMAGUCHI, T. and MAKI, I., *Kogyo Kagaku Zasshi*, **66**, 428 (1963).
[452] INOUE, E., KOKADO, H. and YAMAGUCHI, T., *J. phys. Chem.*, **69**, 767 (1965).
[453] LENDVAY, E., *J. phys. Chem.*, **69**, 738 (1965).
[453a] GILMAN, JR., P. B., *Photogr. Sci. Eng.*, **11**, 222 (1967).
[454] WEST, W., *Photographic Science*, Symposium Zürich 1961, edited by W. F. BERG, p. 71, Focal Press (London) 1963.
[455] NEEDLER, W., GRIFFITH, R. and WEST, W., *Nature, Lond.*, **191**, 902 (1961).
[456] HOLMOGOROV, V. and AKIMOV, I., *Dokl. Akad. Nauk SSSR*, **144**, 402 (1962).

[457] BARANOV, E. V., HOLMOGOROV, V. and TERENIN, A. N., *Dokl. Akad. Nauk SSSR*, **146**, 125 (1962).
[457a] LU VALLE, J. E., LEIFER, A., KORAL, M. and COLLINS, M., *J. Phys. Chem.*, **67**, 2,635 (1963).
[458] OKAMOTO, H. and BRENNER, W., *Organic Semiconductors*, Reinhold Publishing Corporation (New York) 1964.
[459] MEIER, H., *Angew. Chem.*, **77**, 633 (1965); *Angew. Chem. Int. Ed.*, **4**, 619 (1965).
[460] HAMANN, C., *Phys. stat. sol.*, **12**, 483 (1965).
[461] GREIG, H. G., *R.C.A. Rev.*, **23**, 413 (1962).
[462] U.S.P. 3,244,517 (Cl. 96-1) E. LIND (to Azoplate Corp.) April 5 (1966).
[463] U.S.P. 3,122,435, dated 21.4.1961; D.A.S. 1,194,255 Cl. 57b dated 13.7.1960, issued 3.6.1965, R. J. NOE and R. F. HEYLEN (to Gevaert Photo-Producten N.V., Mortsel).
[464] MULDER, B. J. and JONGE, J. DE., *Proc. Kon. nederl. Akad. Wetensch.*, Ser. B, **66**, 320 (1963).
[465] MULDER, B. J., *Philips tech. Rdsch.*, **25**, 195 (1963/64).
[466] STEKETEE, J. W. and JONGE, J. DE, *Proc. Kon. nederl. Akad. Wetensch.*, Ser B, **66**, 76 (1963).
[467] MYLNIKOV, V. and TERENIN, A., *Mol. Phys.*, **8**, 387 (1964).
[468] MYLNIKOV, V. and TERENIN, A., *Dokl. Akad. Nauk SSSR*, **155**, 1167 (1964).
[469] NELSON, R. C., *J. chem. Phys.*, **39**, 112 (1963).
[470] FUJIMORI, E., *Nature, Lond.*, **201**, 1183 (1964).
[471] D.F.P. 1,068,115, H. HOEGL, O. SÜS and W. NEUGEBAUER (1957).
[472] OKA, S., MORI, T., KUSABAYASHI, S., TANIGUCHI, A., YAMAMOTO, Y., JSHIGURO, M. and MIKAWA, H., *Mem. Inst. sci. ind. Res., Osaka Univ.*, 21, **169** (1964).
[473] Belg. P. 588,660, A. G. KALLE (1959).
[474] Brit. P. 975,464, A. G. KALLE (1964).
[475] U.S.P. 3,140,946, P. M. CASSIERS and J. F. WILLIAMS (1964).
[476] SZENT-GYÖRGYI, A., *Science*, **93**, 609 (1941); *Nature, Lond.*, **157**, 9 (1941); *Radiat. Res. Suppl.*, **2**, 4 (1960).

V. THE PHOTOCONDUCTIVITY OF SENSITIZING DYES

5.1 *The Importance of Research Work on the Photoconductivity of Dyes in Relation to Spectral Sensitization*

The electron transfer theory of spectral sensitization (see Chap. VI) postulates the formation and migration of electronic charge carriers in an exposed sensitizing dye. Since, in all the work which had been done on this problem up to the beginning of the fifties, either the idea of photoelectric conductivity in dyes had been rejected,[36,645] or else only a small change in conductivity had been observed as the result of the photochemical decomposition of the dye molecules with long exposures,[477–480] the electron transfer theory gained no recognition.[36,266,310] Even the crystal photoeffect observed in dyes by *Terenin*, *Putzeiko et al.*[251,391,439,481,482] using *Bergmann's* condenser method was not regarded as a proof of the formation of electronic charge carriers in a sensitizer[266] since the insulated electrodes used did not permit the observation of a steady flow of current.

As shown by the results given in this section this view which, among other things led to the rejection of the electronic mechanism of energy transfer must nevertheless have been revised. The numerous investigations which have been carried out since 1951 have given positive proof that sensitizing dyes, on exposure to visible radiation, exhibit photoconductivity which is due to the formation of electronic charge carriers: cf.[53,269,416,459,483–492,643,644] This might furnish the primary prerequisite for the working of the electronic mechanism of sensitization.

A few observations made during the systematic testing of the photoconductivity of dyes make it natural to ask whether the latter is in fact connected with the sensitization reaction:

Firstly the photoconductivity of sensitizing dyes certainly cannot be adduced on its own as a proof that charge carriers are interchanged between the dye and the silver halide. Secondly the fact that desensitizers, photographically inert dyes, aromatic hydrocarbons, high polymers and other organic solids[458,492a,493,494,460] are photoconducting appears to give evidence against a combination of the two effects. A similar construction could also be placed on the absence of any correlation between sensitizing power and the order of magnitude of the photoconductivity of a sensitizer.

When discussing this question the following points must however also be taken into consideration.

1. For the spectral sensitization of the photoconductivity of inorganic semiconductors not only can photographic sensitizing dyes be used but, in addition to desensitizers, photographically inert organic dyes can also be employed

(see Chap. IV, 2.2). This means that since a large number of dyes cannot be used for photographic sensitization, the latter represents a special case among the numerous sensitization reactions in existence. Certain adsorption conditions given by the silver halide grains and also a number of other factors (desensitization, destruction of the dye by bromine, etc.) which are inimical to the photochemical process of latent image formation may possibly be responsible for this state of affairs. It can therefore very probably be assumed:

> that every dye is suitable for the spectral sensitization of a photoconductor if it is capable of being strongly adsorbed to the photoconductor and an interchange of charge carriers is physically possible. This assumption is supplemented by a rule, which has been derived from measurements of the conductivity of a large number of dyes, to the effect that dyes with conjugated double bonds exhibit a photoelectric effect when placed in a suitable measuring instrument.

Both these rules can be taken to indicate that the existence of a connection between the photoconductivity of organic dyes and their sensitizing action cannot be ruled out as a matter of course.

2. In section 2.3 of Chap. VI relations were shown to exist between a number of laws governing the light- and dark-conductivity of dyes and the corresponding laws governing the photoconductivity of dye-sensitized semiconductors and of spectral sensitization respectively. The electrical behaviour of sensitizing dyes consequently has an effect on the sensitization reaction, as can be expected in the event of the two effects being combined.

3. When interpreting the sensitization reaction, the sensitizing power of thick layers of dye of up to the order of magnitude of $1\,\mu$ must be taken into consideration, The resonance hypothesis does not cover this effect which would however be explicable if a relation existed between photoconductivity and sensitization, since photoelectric measurements prove that electronic charge carriers can still migrate through dye crystals of 0·1 mm in thickness and dye layers of up to 1 mm in thickness.[494a]

4. Apart from the reasons given in 1–3, an investigation of the dark- and photo-conductivity of sensitizing dyes is desirable as it might possibly give some insight into the mechanism of sensitization. In this connection the reader is reminded that conclusions regarding the mechanism of heterogeneous catalysis can also be drawn from photoconductivity measurements.[495,496] This approach appears to be particularly promising in the case of supersensitization which is closely bound up with sensitization, since in the former reaction as in that of heterogeneous catalysis a very much greater effect is achieved by the combined action of two substances (in the nature of doping) than can be expected from a summation of the individual actions of the substances.

5. Any discussion of the electron transfer theory of spectral sensitization presupposes a knowledge of the absolute position of the energy terms of the

dye adsorbed to a AgBr grain. Since very little is known as yet about these quantities in the case of organic dyes, and as moreover they can be determined from conductivity and other measurements, this provides a further argument in favour of investigating the conductivity behaviour of organic sensitizers.

5.2 The Measurement of the Photoconductivity of a Dye

5.2.1 METHODS OF MEASUREMENT. In theory, the photoelectric behaviour of organic dyes can be measured in a similar way to that of an inorganic system. Nevertheless the high electrical resistance of (undoped) solid dyes (specific resistance of up to $10^{14}\,\Omega\cdot\text{cm}$), their high extinction, the difficulty of preparing single crystals and the fact that disturbances by means of photochemical

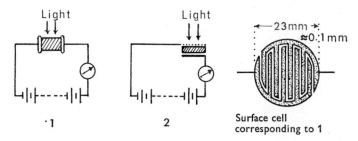

Fig. 20. Diagram of arrangements for measuring the photoconductivity of dyes. 1. Surface type arrangement. 2. Sandwich type arrangement.

reactions often make it necessary to modify the methods which are commonly used in studies of inorganic photoconductors; see the review in[459]. In addition to the method of testing dyes by the Dember effect,[441,481,482,497] the investigation of dye layers by means of the longitudinal field and transverse field cells, or sandwich and surface cells shown diagrammatically in Fig. 20 has proved to be a particularly suitable method.[484-486,492,498,499,644] The Becquerel intrument is used for measuring photochemical dye bleaching-out reactions,[500-504,643] a subject which will not be discussed any further here.

5.2.2 PREPARATION. As already mentioned, dyes are usually studied in thin layers. These are obtained either by rapid precipitation from an easily evaporated solvent or by sublimation in a high vacuum at 10^{-5} torr.[644,488,505-509] It is worth noting that the results obtained with single crystals (including tests on the dependence of conductivity on direction[510-512]) are in accordance with those obtained with layers: this can be seen above all from the measurements made with phthalocyanines.[513-516]

The purification of dyes by chromatography[488,490] or by sublimation in a high vacuum[491,492] does not diminish their photoconductivity: cf. the examples given in Figs. 21 and 22. Since particularly uniform layers can be obtained by sublimation, the photoconductivity of the dyes obtained by this method of preparation is often higher even than that of the dyes obtained by precipitation from a solvent.

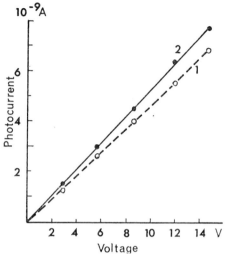

Fig. 21. The influence of chromatographic purification. Hofmann's violet: (1) purified; (2) unpurified; photocurrent as a function of the voltage (transverse field or surface cell).

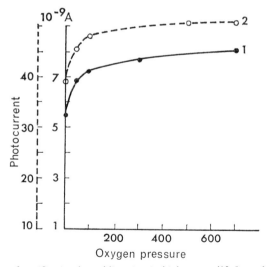

Fig. 22. The influence of purification by sublimation in high vacuo (10^{-5} mm Hg). Merocyanine A10 [59]: (1) unpurified; (2) purified. The photocurrent as a function of the oxygen pressure.

These observations present an argument against the frequently discussed assumption that impurities are ultimately responsible for the photoconductivity of dyes. In this connection it must also be mentioned that the photoconductivity of organic compounds is not removed by purifying them by zone refining.[517]

5.3 A Summary of some of the Laws Governing the Dark- and Photo-conductivity of Dyes

5.3.1 DARK-CONDUCTIVITY. The dark-conductance of dyes which is of the order of magnitude of 10^{-10} to 10^{-14} (Ω^{-1} cm^{-1}) in analogy to inorganic semi-conductors, increases with rise in temperature in accordance with equation 42[53,459,518–524]

$$\sigma_D = \sigma_{D_0} \exp(-\Delta E_{therm}/2\,kT) \qquad (42)$$

where σ_D is the dark-conductivity, σ_{D_0} the pre-exponential factor, k, the Boltzmann constant and T, the absolute temperature.

An interesting fact is that the thermal activation energy ΔE_{therm}, which can be estimated from the slope of the straight lines $\log \sigma_D - (1/T)$ (cf. the example given in Fig. 23) is closely related to the structure of the organic compounds.

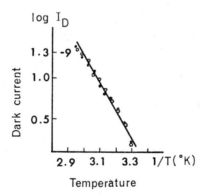

Fig. 23. Pinacyanol, relation between the dark current and the temperature. •: 25° → 80°C; o: 80° → 25°C.

This may be regarded as a proof that the conductivity of dyes and other organic solids is due to the $\pi - \pi^*$-excitation of the organic molecules and not to the action of undefinable impurities.

On the one hand the ΔE_{therm} of linearly anellated hydrocarbons can be stated as a function of the number of rings n by means of the relation (43).[525]

$$\Delta E_{therm} = \frac{(17-n)^2}{100} \qquad (43)$$

On the other hand ΔE_{therm} is generally found to decrease with increase in the number N of the π-electrons, in accordance with the relation given by the electron gas theory.[122,523] If the relations existing between ΔE and N for an open (equations 44a and 44b) and a ring π-electron system (equations 45a and 45b) are represented in the form of a graph—cf. Fig. 24 and Chapter II—then the values of ΔE_{therm} obtained from the conductivity measurements of many organic compounds[526] including those of polymers[517] and dyes[53,528] will lie between the two curves.

Open system

$$\Delta E = \frac{h^2}{8m\,L^2}(N+1)\;[\text{eV}] \qquad (44a)$$

$$\Delta E = 19 \cdot 2\,\frac{N+1}{N^2}\;[\text{eV}] \qquad (44b)$$

Ring system

$$\Delta E = \frac{h^2}{4m\,L^2}\cdot N\;[\text{eV}] \qquad (45a)$$

$$\Delta E = 38\cdot\frac{1}{N}\;[\text{eV}] \qquad (45b)$$

where N is the number of π-electrons; L is the length of the chain and the circumference of the ring, respectively; $L = 1 \cdot N$, whereby $1 = 1\cdot39$ Å is the distance between the C atoms in aromatic compounds.

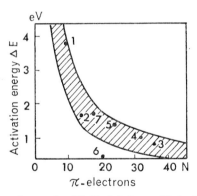

Fig. 24. $\Delta E = f$ (number of π-electrons). (1) Naphthalene, (2) Anthracene, (3) Isoviolanthrone, (4) Pyranthrone, (5) Pinacyanol, (6) α-diphenyl-β-picrylhydrazil, (7) Phthalocyanine.

The last-named relation proves that, to a close approximation, the electron gas theory which was developed for single molecules is applicable to the conduction processes in organic solids (dyes, aromatic compounds). This finding might be regarded of importance, not so much as in enabling the activation energy ΔE_{therm} of the conduction process in organic compounds to be estimated in advance, as in being able to conclude from the anomalies in behaviour, that changes had taken place in the conduction process—migration of protons, etc.[528]—or that the number of mobile π-electrons had decreased.

The above condition is not fulfilled for instance by the α-diphenyl-β-picrylhydrazyl radical when there are 20 π-electrons with $\Delta E_{therm} = 0.26$ (eV), thus confirming the existence of an anomaly in the mechanism of the conductivity in this radical. From the experimental values of $\Delta E_{therm} = 1.71$ (eV) obtained in high vacuo for single crystals of phythalocyanine it can be concluded that an electron system consisting of 18 π-electrons is responsible for the absorption of light and for the conductivity.[53,459]

The last named relation which is indicative of a close connection between the thermal and optical excitation of charge carriers in dyes is also confirmed by the conformity of the values of ΔE_{therm} with those of the energies which can be estimated from the long wavelength limits of the photoconductivity $\lambda_g{}^*$. Cf.

$$\Delta E_{opt} = \frac{h \cdot c}{\lambda_g{}^*} \tag{46}$$

Table 10 and[518,529,530]

TABLE 10

VALUES OF ΔE_{therm} AND ΔE_{opt} OF A FEW DYES DETERMINED FROM EQUATIONS 42 AND 46

	Phenosafranine (*desensitizer*)	*Pinacyanol* (*sensitizer*)	*Cu-Phthalocyanine*
ΔE_{therm}	2·08	1·8	1·7
ΔE_{opt}	2·07	1·77	1·63

5.3.2 PHOTOCONDUCTIVITY IN RELATION TO THE SPECTRUM. Numerous investigations have confirmed that the action spectrum of the photoconductivity measured in thin layers of dye differs little from the absorption spectrum; cf. for example[53,441,644,482,498,499,516,519–532]

This proves that the energy absorbed by the dye molecule—and not by the lattice defects—is responsible for the occurrence of photoconduction. It is worth noting that the bands of associated polymer dyes (*J-bands*) are also perceived in the spectral curves of the photoconductance, as shown by measurements made with the sensitizers diethylpseudoisocyanine and N,N'-diethyl-2,2'-α-naphthothiazole carbocyanine.[533]

The anomalies which can be observed in thick, strongly absorbing layers of dye[498,531] and which occur in surface type cells on irradiating them from the side on which the dye is coated (minimum current strength at about the maximum absorption) might be due to the fact that only part of the superficially formed charge carriers reach the region of the electrical field between the electrodes. When however irradiation is carried out in the reverse direction through the space between the electrodes in these surface cells, these anomalies are therefore removed.

In the sandwich type arrangement, e.g. with single phthalocyanine crystals[516] the action and absorption spectra are however in good agreement. It should be mentioned that this behaviour of single crystals of aromatic compounds can be made use of for determining the absorption spectrum.[534]

5.3.3 PHOTOCONDUCTIVITY IN RELATION TO THE VOLTAGE. Like inorganic systems, dyes and other organic semiconductors are observed to obey Ohm's law, i.e. the dark- or photo-current is proportional to the voltage, generally up to field strengths of 10^3 V/cm[484,489,514,522,644] and not exceeding 10^5 V/cm[526,535] (see the example given in Fig. 25).

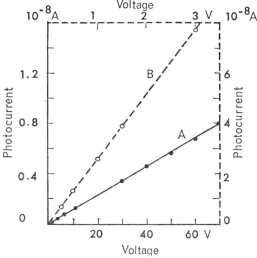

Fig. 25. The photocurrent as a function of the voltage. A: Cu-Phthalocyanine (surface type cell); B: Merocyanine A10 [59] (sandwich type cell). Exposure conditions A: unfiltered light (4000 — 8000 Å), $1·8 . 10^3$ μW/cm². B: $\lambda = 6220$ Å, $370 . 10^{13}$ quanta/cm² sec. Distance between the electrodes A: 0·01 cm, B: 10^{-4} cm. Exposure area: 1 cm².

At higher field strengths the following anomalies occur:

1. The dark- and photo-currents increase more than proportionally with increase in the voltage[53,514,525,536] as exemplified by the infra-red sensitizer neocyanine shown in Fig. 26.

Various hypotheses have been advanced in explanation of this phenomenon which is known as the Wien-Poole effect,[525,537,538] and a review of them is given in[53]. These hypotheses bring the effect partly into relation with a special conduction mechanism which occurs in organic compounds. It might however be more realistic to attribute the effect to the formation of space charge limited currents in organic semiconductors which are characterized by a relatively large distance between their energy bands. According to *Eley*,[526] in the presence of charge carriers which subsequently furnish the ohm contacts, Ohm's relation ($I \sim U$) in the case of space charge limited currents (SCLC)

$$U_E = n_i e \cdot L^2 / 2\varepsilon \qquad (47)$$

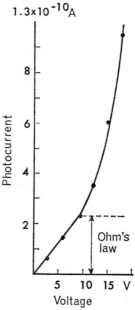

Fig. 26. Neocyanine. The photocurrent as a function of the voltage. $\lambda = 5980$ A, $330 \cdot 10^{13}$ quanta/cm² sec. Distance between the electrodes: 0·01 cm.

(n_i is the density of the thermally excited charge carriers (and the intrinsic conduction density, respectively); L is the distance between the electrodes; ε is the dielectric constant)

is, from a limiting voltage U_E onwards, replaced by a relation which follows approximately the laws derived by *Child*[539] for space charge limited currents in ionized gases:

$$I = \text{const} \cdot \varepsilon \cdot \mu_e \cdot \frac{1}{L^3} \cdot U^2 \quad [A \cdot \text{cm}^{-2}] \qquad (48)$$

When traps are present the mobility μ_e deteriorates, and according to *Eley*,[625] the dependence on voltage is raised from $I \sim U^2$ to $I \sim e^{\alpha U}$. The change in the current-voltage characteristic from $I \sim U$ to $I \sim U^2$ or to $I \sim e^{\alpha U}$, which is observed with various dye-photoconductors from a certain limiting voltage onwards, can therefore be explained on similar lines to the change observed in inorganic semiconductors, so that, on the basis of the observed current-voltage characteristics, no special conduction mechanism has to be argued for dye photoconductors.

In this connection it should be remarked that the use of the techniques of space-charge-limited currents (SCLC)—that is a complete measurement of the current as a function of voltage, temperature, layer thickness and illumination—can give values for the mobility of charge carriers and for the determination of

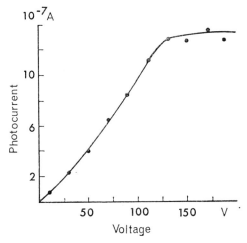

Fig. 27. Crystal violet. The photocurrent as a function of the voltage. $\lambda \times 5870$ Å, $31 . 10^{12}$ quanta/cm² sec. Distance between the electrodes: 0·01 cm.

the properties of traps (e.g. total trap density, trap distribution, trap capture cross section) which are located in the forbidden gap of many compounds.[492a,536,539a-d]

2. With a number of dyes, saturation of the photocurrent occurs at higher field strengths;[484,486] cf. Fig. 27.

Measurement of the saturation current is mainly of importance for determining the quantum yield, since it provides a means of obtaining data of the number of charge carriers formed in a dye on exposure; cf. however[459, 539e,540].

It should be noted that many questions still remain open in connection with the current-voltage characteristics of organic semiconductors: see the discussions in[459] and [492a] (page 568). A systematic testing of metal/dye systems, taking into account the work functions, the nature of the majority carriers, the type of contact and the problem of the space charge limited currents[541,542] would therefore be of interest.

5.3.4 PHOTOCONDUCTIVITY IN RELATION TO THE INTENSITY OF THE IRRADIATION.

A linear increase similar to that which is known to occur in inorganic photoconductors could also be detected in the photoconductivity of numerous dyes in a sandwich and a surface type cell on increasing the intensity of the irradiation.[537,543] The measured values of the I_{phot} — irradiation relation of the sensitizer Pinaverdol are given as an example in Fig. 28. Cf. moreover[490,492,644].

5.3.5 THE QUANTUM YIELD.

The quantum yield G—gain factor according to Rose;[544] cf. the discussion in[53]—by which is understood here the ratio of the number of charge carriers (N_e) flowing in the outer circuit and the number of light-quanta (N_q) absorbed in the same space of time, reaches a value of the

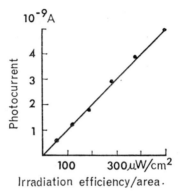

Fig. 28. Pinaverdol. The photocurrent as a function of the irradiation intensity/area. $\lambda \times 6200$ Å.

order of magnitude of 1 in the case of a few dyes.[53,459,486,492] For instance for Malachite green $G = 0.2$, for Pinacyanol $G = 0.37$ and for the merocyanine dye A10 (for the formula see Table 8) $G = 0.6$. These values provide evidence against the assumption that the photoconductance of dyes is the result of a photochemical bleaching out process, since the quantum yield of oxidative bleaching-out is about 10^{-3}.[53] The quantum yield G is dependent on the following factors:

1. As shown by the results in Table 11, G increases with decrease in the thickness of the layer; cf.[269,486,501].
It is worth noting that the quantum yield of the sensitized photolysis of an AgBr layer likewise increases with decrease in the thickness of the layer.

2. When the field strength and the thickness of the layer are maintained constant, the quantum yield increases with decrease in the distance between the electrodes. The example given in Table 12 shows a series of measurements which were obtained with surface cells covered with Crystal violet and electrodes placed at different distances from each other.
This observation proves that the series of relations which are valid for the

photoconductivity of a solid—cf. however the table compiled by *Heiland* and *Mollwo*[545] giving the important photoelectric quantities—also hold for organic dye-photoconductors since the measured dependence of the quantum yield can be derived on the basis of these relations:

If $\eta \cdot z$ charge carriers are formed per cm³ and per second—i.e. with a dye of cross-section $B \cdot D$ (cm²), a distance of L (cm) between the electrodes, a volume

TABLE 11

Dye	Method of measurement	Thickness of the layer	G
Pinacyanol	Surface type cell	$7 \cdot 5 \cdot 10^{-6}$ cm	0·3
		$1 \cdot 8 \cdot 10^{-5}$ cm	1·13
		$7 \cdot 5 \cdot 10^{-5}$ cm	0·03
Crystal violet	Surface type cell	$3 \cdot 0 \cdot 10^{-6}$ cm	0·33
		$7 \cdot 0 \cdot 10^{-6}$ cm	0·13
		$1 \cdot 0 \cdot 10^{-5}$ cm	0·06
Cyanine	Becquerel effect	$0 \cdot 35 \cdot 10^{-6}$ cm	$1 \cdot 43 \cdot 10^{-2}$
		$0 \cdot 59 \cdot 10^{-6}$ cm	$0 \cdot 96 \cdot 10^{-2}$
		$0 \cdot 95 \cdot 10^{-6}$ cm	$0 \cdot 99 \cdot 10^{-2}$
		$5 \cdot 74 \cdot 10^{-6}$ cm	$1 \cdot 06 \cdot 10^{-2}$
		$11 \cdot 48 \cdot 10^{-6}$ cm	$0 \cdot 87 \cdot 10^{-2}$
		$17 \cdot 2 \cdot 10^{-6}$ cm	$0 \cdot 72 \cdot 10^{-2}$
		$23 \cdot 0 \cdot 10^{-6}$ cm	$0 \cdot 77 \cdot 10^{-2}$

TABLE 12

CRYSTAL VIOLET $\lambda = 5870$ Å; $E = 700$ V/cm; THICKNESS OF THE LAYER $3 \cdot 10^{-6}$ cm

Distance between the electrodes in cm	N_q	N_e	G
0·08	$177 \cdot 10^{12}$	$1 \cdot 6 \cdot 10^{11}$	$0 \cdot 09 \cdot 10^{-2}$
0·042	$92 \cdot 10^{12}$	$1 \cdot 06 \cdot 10^{11}$	$0 \cdot 12 \cdot 10^{-2}$
0·02	$46 \cdot 10^{12}$	$1 \cdot 2 \cdot 10^{11}$	$0 \cdot 26 \cdot 10^{-2}$
0·01	$25 \cdot 10^{12}$	$1 \cdot 3 \cdot 10^{11}$	$0 \cdot 52 \cdot 10^{-2}$
0·003	$7 \cdot 1 \cdot 10^{12}$	$1 \cdot 0 \cdot 10^{11}$	$1 \cdot 4 \cdot 10^{-2}$

Vl (cm³): $\eta \cdot z \cdot$ Vl charge carriers—and at a given moment these charge carriers have a mean sliding travel (Schubweg) of

$$w = \tau \cdot \mu_e \cdot E \quad (49)$$

(τ: lifetime, μ_e: mobility (cm²/$V \cdot$ sec), E: field strength (V/cm)) then it follows that the photocurrent flowing in the outer circuit when the distance between the electrodes is L will be given by

$$I_{\text{phot}} = \eta \cdot z \cdot \text{Vl} \cdot e \cdot \tau \cdot \mu_e \cdot \frac{E}{L} \qquad (50)$$

z is the number of light-quanta absorbed per cm^3 . sec and η—which is defined as the quantum efficiency or the quantum yield,[544,545] or again as the excitation factor[53]—is the fraction of the charge carriers which are primarily formed from the light-quanta absorbed (limiting value $\eta \to 1$). Since the quantum yield G (gain factor) is defined here as the ratio of the number of charge carriers flowing in the outer circuit ($N_e = I_{\text{phot}}/e$) and the number of the light-quanta absorbed in the same space of time ($N_q = Z \cdot \text{Vl}$) (equation 51)

$$G = \frac{I_{\text{phot}}/e}{z \cdot \text{Vl}}, \qquad (51)$$

the observed relation follows from equation 50:

$$G = \eta \cdot (\tau \cdot \mu_e) \cdot E \cdot \frac{1}{L} \qquad (52)$$

It should be mentioned that equation 52 was also made use of for determining the sliding travel $w(= \tau \cdot \mu_e \cdot E)$ of dyes in the case in which $\eta \to 1$; cf. [53,486]

3. According to previous measurements the following relation holds between G and the structure of a dye:

a. In triphenylmethane dyes the degree of methylation of the amino groups is proportional to the photoconductivity. Replacement of a methyl group reduces the sensitivity in the same way as replacement by hydrogen: that is to say:

$$CH_3 \gg C_2H_5, H, C_6H_5$$

b. As measurements of cyanines and merocyanines have proved, the photoconductance of dyes increases with the number of methine groups, i.e. with increase in the length of the conjugated chain.

These structural relations can be regarded as a proof that the photoconductivity of dyes is not due to undefinable impurities.

5.3.6 THE INERTIA OF THE GROWTH AND DECAY PROCESS. A photocurrent develops within 10^{-5} to 10^{-6} sec in dyes, as has been shown by measurements made with a Cu-phthalocyanine and a phthalocyanine not containing a metal, etc.;[498,546] cf. also [484,499]. It has not yet been possible to represent this increase in the photocurrent by means of a simple time-law, possibly because the current develops too rapidly to be measured by any of the apparatus which has been available up to the present.

It should be noted that the rapid formation of the photocurrent in combination with the thermal stability and high resistance of Cu-phthalocyanine and other organic semiconductors enable them to be used in Vidikon television receivers.[53,547]

At the end of the exposure the current, with time constants ranging from 10^{-3} to at least several seconds, decreases. As shown by the example given in Fig. 29, this drop in current can be expressed by equation 53:

$$\frac{1}{I(t)} = k \cdot t + \text{const} \qquad (53)$$

i.e. the process of recombination is a bimolecular one.[518]

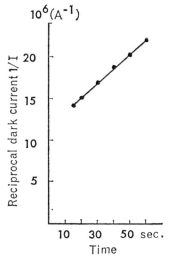

Fig. 29. Pinacyanol. Decrease in the dark-current (after the end of the exposure) as a function of the time.

With large photocurrents, corresponding to a high concentration of carriers, a slower drop is observed than that which can be expected from equation 53. Whereas the argument put forward by *Tollin et al.*[498] to explain this anomalous decrease in current is to the effect that the mobility of the charge carriers is dependent on their concentration, *Waite*[548] on the other hand assumes that the bimolecular recombination is strongly influenced by diffusion processes. A few analogies between inorganic and organic photoresistances suggest however that traps also have an influence on the sluggish decrease in the photocurrent.[53,459,530,549]

It was found for example that the inertia effects in dyes could be greatly decreased by means of high field strengths and the use of cells with electrodes placed close together, as is characteristic of photoconductors containing traps.[544,550] Moreover the charge carriers present at the end of the period of exposure can be "illuminated" (glow curves) by raising the temperature both of dyes[549] and of other organic photoconductors,[551–553] as is known to be the case with inorganic crystal phosphors which are rich in traps.[554,555,556] Again, other temperature effects of dye photoconductance also indicate the presence of traps (see the next section).

It should be noted that the discussions on the use of organic photoconductors for the storage of energy (photoelectrets) are based on the effects which are observed to be produced by the traps.[557]

5.3.7 TEMPERATURE EFFECTS. The photocurrent increases with increase in temperature in accordance with equation 54.

$$\sigma_{\text{phot}} = \sigma_0 \cdot \exp\left(-\Delta E_H / kT\right) \tag{54}$$

Values of σ_{phot} plotted against $1/T$ therefore give a straight line (see Fig. 30) from the slope of which the thermal activation energy ΔE_H of the photoconductance can be calculated.[549] As Table 13 shows, ΔE_H is of the order of

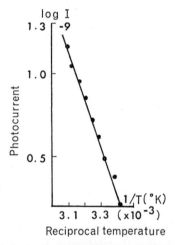

Fig. 30. Pinacyanol. Relation between the photocurrent and temperature. Field strength 600 V/cm; $\lambda = 5900$ Å.

0·4 to 0·6 eV and is not identical with ΔE_{therm} (e.g. for Orthochrome T: $\Delta E_{\text{therm}} = 1\cdot5$ eV; $\Delta E_H = 0\cdot44$ eV).

In analogy to crystal phosphors[556,559,560] the dependence of the photocurrent on temperature might be attributable to the action of traps which are filled by some of the charge carriers formed during exposure, because these lattice defects are emptied on raising the temperature, thus increasing the photocurrent. The change in the rate constant of the drop in the photocurrent observed on raising the temperature is in accordance with this hypothesis[549] as is also the fact that "photoelectric glow curves"[549,561] can be measured.

It is therefore natural to attribute the trapping states to defects in the lattice structure of the molecules of the organic layers studied, i.e. to assume the presence of impurities, irregular lattice structures, dislocations or vacant sites. It must however be noted that the utmost careful chromatographic purification or repeated sublimation in a high vacuum fail to produce any effect on the phenomena due to secondary effects. The effects are not limited to

TABLE 13

Dye	ΔE_H(eV)	Note
Pinacyanol	0·56	
Orthochrome T	0·44	
Rhodamine B	0·62	0·54 eV according to[558]
Crystal violet	0·48	
Methylene blue	0·68	
Brilliant green	0·58	
Eosine	0·46	

powders which promote the formation of lattice defects, but they can also be perceived in single crystals.

This may possibly be regarded as an indication of the influence of terms belonging specifically to dyes and organic photoconductors, respectively, these terms acting in a similar way to traps. It may still be uncertain to what extent these terms may be identified with the triplet states which lie below the singlet levels—see review in[562]—and which cannot usually be detected in solid dyes on account of radiationless deactivation processes.[562a] In this connection the fact that in vitreous, set melts of dye and boric acid it is possible to fix the excited electrons after the π–π* excitation secondarily to the triplet terms (traps) in the dye, might be of significance; cf. also the corresponding discussion in[53], p. 238.

5.3.8 *n*- AND *p*-CONDUCTION IN DYES. On the basis of the changes which can be observed in the dark- and photo-conductivity of dyes when acted upon by oxygen, dyes can be divided into *n*- and *p*-conductors. The conductivity of some dyes (triphenylmethane dyes, rhodamines, etc.) is decreased by the adsorption of oxygen (*n*-conduction), whereas in others (merocyanines, fluoresceins, phthalocyanines, etc.) it is increased (*p*-conduction). Hydrogen has exactly the opposite effect.

Figures 31 and 32 show these effects in the case of an *n*-conducting dye and a *p*-conductor, respectively.

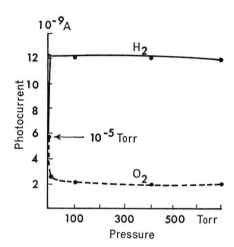

Fig. 31. Parafuchsine. Relation between the photocurrent and the pressure in atmospheres of oxygen and hydrogen. Field strength 500 V/cm (electrode screen cell). $\lambda = 5495$ Å, $157 \cdot 10^{13}$ quanta/sec cm².

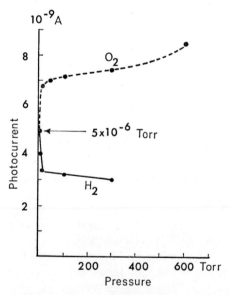

Fig. 32. Merocyanine A10. Relation between the photocurrent and the pressure in atmospheres of oxygen and hydrogen. Field strength; 8000 V/cm (electrode screen cell). $\lambda = 5000$ Å, $120 \cdot 10^{13}$ quanta/sec cm².

The classification of dyes which is derived from the changes in their conductivity assumes:[53,485,487,563]

1. that an electronic interchange, analogous to that which occurs in inorganic semiconductors and metals, also takes place between the adsorbed gas and the solid dye,[564-569] and
2. that the method used by *Heckelsberg, et al.*[570] for estimating the majority carriers is also applicable to organic systems.

Both of these assumptions can now be considered to have been proved. Firstly gases such as NO, BF_3, SO_2 and NH_3, C_2H_5OH, $(CH_3)_3N$, etc. produce the same changes in the conductivity of organic solids as in inorganic

TABLE 14

SUMMARY OF SOME *n*- AND *p*-CONDUCTING DYES.
ESTIMATION BY THE INFLUENCE OF GASES (O_2, H_2)
(For further examples see[53,485])

n-Type of conduction		*p*-Type of conduction	
Dye	Class of dye	Dye	Class of Dye
Fuchsine	Triphenylmethane	Eosine	Fluorescein
Crystal violet	,,	Fluorescein	,,
Hofmann's violet	,,	Erythrosine	,,
Victoria blue	,,	Rose Bengals	,,
Malachite green	,,	Phloxine BBN	,,
Rhodamine B	Phenocarboxonium	Acridine orange	Acridine
Rhodamine S	,,	Acridine yellow	,,
Methylene blue	Phenothiazine	Congo red	Azo dyes
Methylene green	,,	Thioflavine	,,
Pinacyanol	Cyanine dye	Flavine	Isoalloxazine
Orthochrome T	,,	Phthalocyanine	Phthalocyanine
Pinachrome	,,	Cu-, Mg-, Co-Phthalocyanine	,,
Phenosafranine	Phenylphenazonium	Merocyanine A 10	Merocyanine dyes [59]
Safranine T	,,	Merocyanine FX 30	Merocyanine dyes [62]
		Carotine	
		Chlorophyll	

systems.[53,529,571-577] Secondly the *n*- and *p*-mechanisms of conduction in dyes are confirmed by measurements of the *Hall* effect,[513,578,579] of the thermoelectric effect[494,510,580-584] and of the polarity of the diffusion EMF in a static condenser,[394,441] and also by observations of the changes in the contact potential by means of the vibration method.[530]

Moreover the same type of conduction as that found in layers was occasionally established in single crystals. With few exceptions, dyes of one class

133

exhibit the same type of conduction, as can be seen from the survey given in Table 14. For instance p-conduction was observed in more than 40 merocyanine dyes. This proves that the conductivity depends on the structure of the dye and that it cannot be attributed to undefinable impurities. Purification by sublimation or chromatography therefore does nothing to alter the type of conduction.

No answer is yet forthcoming as to what causes the different types of conduction in individual dyes. Experiments which succeed in producing a reproducible change in the type of conduction may possibly provide some contribution to a solution of the problem. For instance the p-conduction of metal

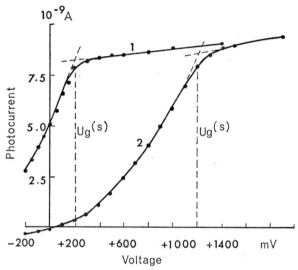

Fig. 33. The external photoelectric current as a function of the voltage. $\lambda = 230$ mμ. (1) Phthalocyanine (p-type). (2) Pinacyanol (n-type).

complexes of phthalocyanine on undergoing a structural change into chlorindium-chlorophthalocyanine[580] is transformed into n-conduction, and a reproducible alteration in the type of conduction can also be achieved by mixing dyes together;[485] see also section 2.2.6. of Chap. VI.

The n- and p-types of conduction in dye-photoconductors also has an effect on the position of the Fermi levels, as can be expected from theory. In the case of exposed dyes of the n-type, $E_F{}^*$ is situated in the neighbourhood of the conduction band, whereas with p-conducting dyes it is displaced towards the fundamental band.[585]

This result was derived from the current voltage characteristics of the external photoelectric effect in various dye-photoconductors (see Fig. 33), for the saturation voltage $U_g{}^{(s)}$ of these characteristics corresponds to the contact potential of the dye/anode (e.g. dye/silver) and therefore enables the thermal work function W_{therm} of the dye, which in a high vacuum can be taken as being approximately equal to the energy of the Fermi level, to be calculated from equation 55.

$$U_g^{(s)} \gtrless 0; \quad W_{\text{dye}} = W_{\text{anode}} \mp eU_g^{(s)} \tag{55}$$

A few of the contact potentials of n- and p-conducting dyes against silver obtained by this method[586] as well as the thermal work functions derived from these results are compiled in Table 15.

It can be seen from the table that the contact potentials of the dyes with n- and p-types of conduction, measured against silver, and therefore the work functions W_{therm} which correspond approximately to the Fermi energies differ

TABLE 15

Dye	Type of conduction	Contact potential against Ag (mV)	W_{therm} (eV)
Phthalocyanine	p	+200	4·5
Merocyanine A 10 [59]	p	+250	4·45
Erythrosine	p	+200	4·5
Pinacyanol	n	+1200	3·5
Malachite green	n	+900	3·8
Orthochrome T	n	+1100	3·6
Phenosafranine	n	+800	3·9

from one another. The energy diagram given in Fig. 34 illustrates that this difference in the energy diagrams of various dye-photoconductors (cf. Chap. VI, section 2.3) takes effect in the said manner: i.e. the Fermi level is displaced towards the conduction band of dyes of the n-type of conduction and towards the valency band of dyes of the p-type of conduction.

5.3.9 THE LIFETIME AND THE MOBILITY OF THE CHARGE CARRIERS. The lifetime τ: it is not yet possible to give an accurate statement regarding the lifetime τ during which the charge carriers are freely mobile in solid dyes, because the curves of the decay in the current which ranges from 10^{-2} seconds to minutes, are in all probability distorted by the periodical accumulation of charge carriers at traps, so that they cannot give any indication of τ.

The assumption that triplet states are primarily involved in the conduction process and that a lifetime of at least 10^{-4} sec has thus to be reckoned with,[498] is ruled out by short-time measurements of phthalocyanines which reveal the formation of charge carriers in less than 10^{-6} sec.[587] Since, on the other hand, the lifetime of excited singlet states in phthalocyanines is about 10^{-9} sec,[588] values of between 10^{-6} and 10^{-9} secs for the lifetime are open to discussion: cf. also[514].

The mobility μ_e: *Delacote* and *Schott*,[579] on the basis of measurements of the *Hall* effect in phthalocyanines, have assigned a value of $\mu_\oplus \approx 100$ cm²/V . sec to the mobility of the charge carriers. This result makes it natural to enquire to what extent the smaller mobilities derived from the ESR investigations[589]

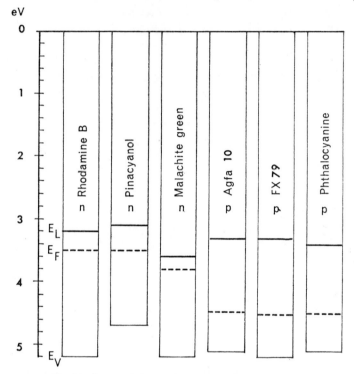

Fig. 34. Energy diagram of *n*- and *p*-conducting dyes.

and from the mean free path lengths[53,490] and again from other measurements[53,529,531,573,590,591] have to be corrected.

A comparison with germanium ($\mu_\oplus = 1700$ cm²/V . sec) is of interest as it shows that the mobilities of the charge carriers in inorganic semiconductors and in organic dyes (at least in a few cases) do not differ altogether too much from one another.

5.4 The Mechanism of the Conductivity of Dyes

The conformity in the electrical behaviour of organic and inorganic semiconductors can be taken as evidence that the photoconductivity of solid organic dyes is not due to photochemical decomposition reactions or to an ionic mechanism but to the formation and migration of electronic charge carriers.

This raises the question as to how an electronic charge carrier is transported through an organic solid. For, whereas the distances between the points in the lattice of an inorganic crystal are about 1·4 Å, the molecules in organic lattices are relatively far apart, namely about 3·5 Å.[53,103,223,592]

Since the problem of the mechanism of the photoconductivity of dyes has already been discussed in detail in[53], only the essential arguments which take the most recent results into account will be summarized in the following. These results argue in favour of a quasi-free transport of the electronic charge carriers on the lines of the band hypothesis.

5.4.1 THE TUNNEL EFFECT MECHANISM. In order to explain the mechanism of the transport of the charge carriers, in various research work[499,525,593] it has been assumed that an excited electron (e.g. in the singlet state $^1E_1$) has made its way via the threshold potential to an unoccupied term of the neighbouring molecule by means of a tunnel effect. Since, according to equation 62, and on the basis of the conditions obtaining in the dye, the probability of this transfer $D.z \approx 10^{14}$ is greater than the probability $(= 1/\tau)$ of a return from the $^1E_1$ state to the ground state, the transport of charge carriers from molecule to molecule by means of the tunnel effect (the so-called "hopping mechanism") appears to be a controversial point. By increasing the ratio which is given in equation 56 and which would occur for instance if a triplet state of long duration $^3E_1$ were to participate in the photoconductivity,[594,595] this mechanism would doubtless be even more probable.

$$\frac{D.z}{1/\tau} = \frac{10^{14}}{10^9} \qquad (56)$$

5.4.2 THE BAND MODEL HYPOTHESIS. The following objections can be raised to the tunnel effect mechanism in which the charge carriers are localized for varying periods of time at the lattice points:

1. Triplet states which would specially favour the mechanism are not primarily involved in the conduction process.[596]
2. A mobility of $\mu \approx 1$ cm^2/V . sec is usually assumed for the limiting value between the "hopping mechanism" and the "quasi-free transport in the sense of the band hypothesis." Since μ-values of up to 100 cm^2/V . sec were measured in the case of phthalocyanines,[579] and the photo-Hall mobility exceeds 50 cm^2/V . sec for electrons and 30 cm^2/V . sec for holes in anthracene,[596a] the energy band hypothesis may be assumed to be the more probable one.
3. An argument in support of the free movement of electrons and positive holes in energy bands is moreover given by the decrease in their mobility

in single crystals of organic compounds on raising the temperature: μ_e was determined by means of an impulse method (light flashes of 1–2 μsec) and using equation 57[459,494a,597,598,598a] and by measurements of the dark-Hall mobility.[598b]

$$\mu_e = \frac{L^2}{U \cdot T} \tag{57}$$

(where L is the distance between the electrodes, U is the voltage, and T is the time of transit measured oscillographically).

If the effect were indeed a tunnel effect an inverse dependence on temperature would have to be observed owing to the increase in the number of impulses z corresponding to equation 56.

What picture can be formed of the manner in which energy bands occur in a solid organic dye?

It can be assumed that on transition to the solid state the discrete energy levels, which can be occupied by electrons, in the dye are split up into many components by electronic interaction. Since the degree of overlapping of the π-electron clouds is ultimately responsible for the differences in energy in the part-levels—i.e. for the width of the band—on the basis of the relatively great distance of about 3·5 Å between the molecules, a band which is narrow in comparison with that of inorganic solids has to enter into the calculations.

The very important part played by the mutual orientation of the molecules in the formation of these energy bands can readily be understood, since any arrangement of flat molecules with their fronts facing one another would not only give rise to high potential thresholds between the molecules, but would even largely decrease the overlapping of the π-electron clouds vibrating above and below the plane of a molecule. A plane to plane arrangement of the dye molecules would, on the other hand, guarantee the maximum possible mixing of the π-electron clouds appertaining to the individual molecules.

A point of great importance in the problem now under discussion is that the transition to the crystalline state often brings the organic molecules precisely into the last-named arrangement:

> The flat molecules stack themselves, not in any random order, but in conformity with a law which is valid in organic chemistry, with their planes on top of one another—just as in a slanting pile of coins—and thus the π-electrons in many organic solids overlap to the greatest possible extent.

This phenomenon very probably represents a general, universally organized principle of the organic process of crystallization which it has not yet been possible to prove in a number of compounds simply because of the overlapping of various groups of aggregates. For instance this mechanism has been proved in the case of the pseudocyanines, the molecular planes of which are arranged at a distance of 3·6 Å from one another at an angle of 59° to the axis of the fibre. Moreover the flat structure of the phthalocyanines[592] which has been confirmed by X-ray investigations, leads one to expect that the molecules which crystallize out in the form of monoclinic prisms will lie on top of each other with their planes parallel and at an angle of 44° to the plane of the axis,

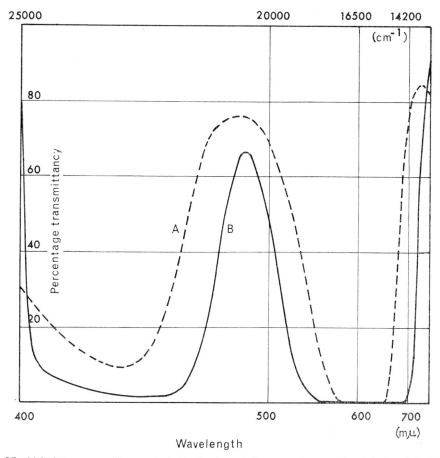

Fig. 35. Malachite green. The spectral distribution of the transmittance. A: Solution (6·7 . 10⁻⁵ mol/lirte). B: Layer (10⁻⁴ cm).

and placed at a distance of 3·5 Å from one another. Even the flat bases of the desoxynucleic acids[539,600] follow this principle; cf. the further examples given in[53].

In addition to the above stated findings, the results of measurements which enable a relation to be recognized between the direction of conductivity and the

Seebeck effect,[597,601-603] as well as the agreement between the spectra of solids and the activation spectra of the photoconduction are in conformity with the band hypothesis.

Moreover the broadening of the bands belonging to the individual molecules observed on passing over to the solid phase is a proof that a splitting up of the specific functions of the molecules by the overlapping of the π-electrons must

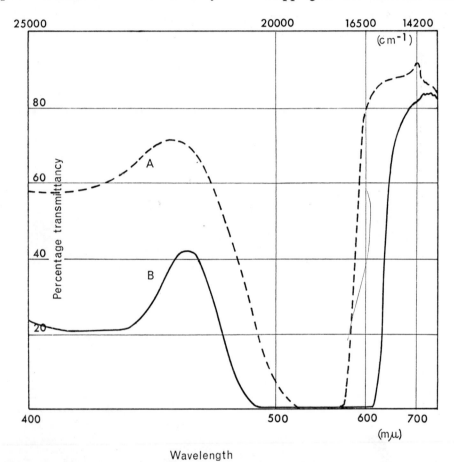

Fig. 36. Rhodamine B. The spectral distribution of the transmittance. A: Solution (4·6 . 10⁻⁵ mol/litre). B: Layer (10⁻⁴ cm).

take place. In other words, on crystallization, the sharp energy levels of the free molecules are transformed into broad, quasi-continuous bands, and this is indicative of mutual interaction between the individual molecules.

Figures 35 and 36 which give the spectra of Malachite green and Rhodamine B in solution and in the solid state as examples show that the transition of a dye from solution to the solid state is accompanied by a broadening of the order of magnitude of $\Delta\tilde{\nu} \approx 2500$ (cm^{-1}) ($\approx 0{\cdot}3$ eV) of the absorption band, which is situated at the longest wavelength.

A change in the vibration structure cannot be held responsible for this broadening since no influencing of the film extinction curve can be detected on cooling to $-180°$C.[53,604]

For information on the problem concerning the narrow paramagnetic electron resonance signals (EPR) of organic semiconductors which exhibit a g-factor of 2·0036 corresponding to a delocalized electron, cf. *Hamann, et al.*[460,605-608]

5.4.3 THE EXCITON HYPOTHESIS. In connection with the photoconductivity of organic compounds various authors[466,531,598,609] discuss the primary formation of excitons (cf. moreover 53) which diffuse on the surface, decompose there, and contribute to the formation of electrons and positive holes. The migration of the electronic charge carriers which are thus formed secondarily are then likewise interpreted by means of the band model.

An argument against the exciton theory is that it does not furnish an explanation of dark-conductivity—see section 3.1 of this chapter. Even in the case of organic dyes which are good conductors and which give a quantum yield G of the order of magnitude of 1 it might not be necessary to assume in addition that only some of the excitons disintegrate into charge carriers. Moreover there are no indications of the presence of any exciton spectra in the case of organic dyes.

References

[477] ZCODRO, N., *Izvest. Akad. Nauk.*, **6**, 727 (1919); *J. Chim. phys.*, **26**, 117 and 178 (1929).
[478] WARTANJAN, A. T., *J. Chim. phys.*, **22**, 769 (1948); **24**, 1361 (1950); *Acta phys. chim., URSS*, **22**, 201 (1947).
[479] PETRIKALN, A., *Z. phys. Chem.*, **B10**, 9 (1930).
[480] NELSON, R. C., *J. chem. Phys.*, **19**, 798 (1951).
[481] PUTZEIKO, E., *J. Chim. phys.*, **22**, 1172 (1948); *Dokl. Akad. Nauk SSSR*, **59**, 471 (1948).
[482] PUTZEIKO, E., *Izvest. Akad. Nauk SSSR, Ser. fiz.*, **13**, 224 (1949); *Abh. soviet Phys.*, **1**, 63 (1951).
[483] NODDACK, W. and MEIER, H., *Z. Elektrochem.*, **57**, 691 (1953).
[484] MEIER, H., *Z. Elektrochem.*, **58**, 859 (1954); *Angew. Chem.*, **68**, 525 (1956).
[485] MEIER, H., *Z. wiss. Photogr.*, **52**, 1 (1958).
[486] NODDACK, W., MEIER, H. and HAUS, A., *Z. phys. Chem.*, **212**, 55 (1959).
[487] MEIER, H., *Z. phys. Chem.*, **212**, 73 (1959).
[488] NODDACK, W., MEIER, H. and HAUS, A., *Z. phys. Chem.* (N.S.), **20**, 233 (1959).
[489] NODDACK, W. and MEIER, H., *Scientific Photography*, Proc. Int. Colloq., Liége 1959, edited by H. SAUVENIER, p. 515, Pergamon Press (Oxford) 1962.
[490] NODDACK, W., MEIER, H. and HAUS, A., *Z. wiss. Photogr.*, **55**, 7 (1961).
[491] MEIER, H., *Kép-és Haugtechnika (Budapest)*, **1**, 1 (1962).
[492] MEIER, H., *Photogr. Sci. Eng.*, **6**, 235 (1962).
[492a] GUTMAN, F. and LYONS, L. E., *Organic Semiconductors*, John Wiley (New York, London, Sidney) 1967.
[493] BROPHY, J. J. and BUTTREY, J. W., *Organic Semiconductors*, Macmillan (New York) 1962.

[494] JUSTER, N. J., *J. chem. Educ.*, **40**, 547 (1963).
[494a] MEIER, H. and ALBRECHT, W., *Z. Elektrochem., Ber. Bunsenges. phys. Chem.*, **71**, 922, (1967).
[495] SCHWAB, G. M., *Angew. Chem.*, **75**, 149 (1963).
[496] HAUFFE, K., *Reaktionen in und an festen Stoffen (Reactions in and on solids)* 2nd edn. Springer-Verlag (Berlin, Heidelberg, New York) 1966.
[497] MEIER, H., *Z. Elektrochem., Ber. Bunsenges. phys. Chem.*, **59**, 1029 (1955).
[498] TOLLIN, G., KEARNS, D. R. and CALVIN, M., *J. chem. Phys.*, **32**, 1013 (1960).
[499] HAUFFE, K. and KAUFHOLD, J., *Z. Elektrochem., Ber. Bunsenges. phys. Chem.*, **66**, 316 (1962).
[500] INOKUCHI, H., MARUYAMA, Y. and AKAMATU, H., *Proceedings of the Conference on Electrical Conductivity in Organic Solids*, edited by H. KALLMANN and M. SILVER, p. 69, Interscience (New York) 1961.
[501] ECKERT, G., *Z. wiss. Photogr.*, **49**, 220 (1954).
[502] HILLSON, P. J. and RIDEAL, E. K., *Proc. roy. Soc.*, Ser. A. **216**, 458 (1953).
[503] COPELAND, H. W., BLACK, O. D. and GARRETT, A. B., *Chem. Rev.*, **31**, 177 (1942).
[504] EVSTIGNEEV, V. B., SAVKINA, I. G. and GAVRILOVA, V. A., *Biofizika*, **7**, 298 (1962).
[505] NELSON, R. C., *J. molecular Spectroscopy*, **7**, 449 (1961).
[506] ELEY, D. D., PARFITT, G. D., PERRY, M. J. and TAYSUM, D. H., *Trans. Faraday Soc.*, **49**, 79 (1953).
[507] WIHKSNE, K. and NEWKIRK, A. E., *J. chem. Phys.*, **34**, 2184 (1961).
[508] LIANG, C. Y. and SCALCO, E. G., *J. electrochem. Soc.*, **110**, 779 (1963).
[509] DANZIG, M. J., LIANG, C. Y. and PASSAGLIA, E., *J. Amer. chem. Soc.*, **85**, 668 (1963).
[510] FIELDING, P. E. and GUTMAN, F., *J. chem. Phys.*, **26**, 411 (1956).
[511] METTE, H. and PICK, H., *Z. Phys.*, **134**, 566 (1953).
[512] RIEHL, N., *J. phys. Chem.*, **29**, 459 (1955).
[513] HEILMEIER, G. H., WARFIELD, G. and HARRISON, S. E., *Phys. Rev. Letters*, **8**, 309 (1962).
[514] HEILMEIER, G. H. and WARFIELD, G., *J. chem. Phys.*, **38**, 163 (1963).
[515] BOWDEN, F. P. and CHADDERTON, L. T., *Proc. roy. Soc.*, Ser. A. **269**, 143 (1962).
[516] CADDERTON, L. T., *J. phys. Chem. Solids*, **24**, 751 (1963).
[517] AFTERGUT, S. and BROWN, G. P., *Organic Semiconductors*, edited by J. J. BROPHY and J. W. BUTTREY, p. 79, Macmillan (New York) 1962.
[518] MEIER, H., *Z. phys. Chem.*, **208**, 325 (1958).
[519] VARTANYAN, A. T. and KARPOVICH, I. A., *Dokl. Akad. Nauk SSSR*, **111**, 561 (1956); **113**, 1020 (1957).
[520] VARTANYAN, A. T. and ROSENSHTEIN, L. D., *Dokl. Akad. Nauk SSSR*, **124**, 195 (1959).
[521] VARTANYAN, A. T., *Izvest. Akad. Nauk SSSR, Ser. fiz.*, **20**, 1541 (1956); **21**, 523 (1957); *Soviet Phys. Ber.*, **7**, 332 (1962).
[522] KLEITMAN, D. and GOULDSMITH, G. J., *Phys. Rev.*, **98**, 1544 (1955).
[523] ELEY, D. D. and PARFITT, G. D., *Trans. Faraday Soc.*, **51**, 1529 (1955).
[524] HAMANN, C., *Berichte aus dem Institut für angewandte Physik der Reinstoffe (Reports of the Institute of the Applied Physics of Pure Substances* (Dresden) IfR.-Nr. 13, November, 1962.
[525] WILK, M., *Z. Elektrochem.*, **64**, 930 (1960).
[526] ELEY, D. D., *Semiconductivity and photoconductivity in Organic Substances*, in *Optische Anregung organischer Systeme (Optical Excitation of Organic Systems)* (edited by W. FOERST) Verlag Chemie, Weinheim/Bergstr., pp. 600–633, 1966.
[527] POHL, H. A. and CHARTOFF, R. P., *J. Polymer. Sci. A*, **2**, 2787 (1964).
[528] POLLOCK, C. MC. and UBBELOHDE, A. R., *Trans. Faraday Soc.*, **52**, 1112 (1956).
[529] FOX, D., LABES, M. M. and WEISSBERGER, A., *Physics and Chemistry of the Organic Solid State*, Vol. II. Interscience Pub. (New York) 1965.
[530] TERENIN, A., *Electrical Conductivity in Organic Solids*, H. KALLMANN and M. SILVER, p. 39, Interscience Pub. (New York) 1961; *Proc. chem. Soc., Lond.*, 321, 1961.
[531] WEIGL, J. W., *J. chem. Phys.*, **24**, 883 (1956); *J. mol. Spectr.*, **1**, 216 (1957).
[532] DEVEAUX, P., SCHOTT, M. and LAZERGES, M., *Phys. stat. sol.*, **4**, 43 (1964).
[533] DÖRR, F. and KLEINLE, A., *Z. Naturf.*, **13b**, 257 (1958).
[534] CARSWELL, D. J., *J. chem. Phys.*, **21**, 1890 (1953).
[535] NORTHROP, D. C., *Proc. phys. Soc.*, Sect. B, **74**, 756 (1959).
[536] ROSE, A., *Phys. Rev.*, **97**, 1538 (1955).
[537] JOFFE, A. F., *Physik der Halbleiter (Physics of Semiconductors)* Akademie Verlag (Berlin) 1958.

[538] RIEHL, N., *Ann. Phys.*, (6), **23**, 90 (1957).
[539] CHILD, C. D., in *Handbuch der Experimentalphysik* (*Manual of Experimental Physics*), Vol. 13, Part 1, p. 92, Akad. Verlagsges. (Leipzig) 1929.
[539a] LAMPERT, M. A., *Rept. Progr. Phys.*, **27**, 329 (1964); *Phys. Rev.*, **103**, 1,648 (1956).
[539b] JANSEN, P., HELFRICH, W. and RIEHL, N., *Phys. Stat. Sol.*, **7**, 851 (1964).
[539c] WEISZ, S. Z., COBAS, A., RICHARDSON, P. E., SZMANT, H. H. and TRESTER, S., *J. chem. Phys.*, **44**, 1,364 (1966).
[539d] SUSSMAN, A., *J. appl. Phys.*, **38**, 2,738, 2,748 (1967).
[539e] STÖCKMANN, F., *Z. Phys.*, **138**, 404 (1954).
[540] POLKE, M., STORCH, G. and STÖCKMANN, F., *Z. Phys.*, **55**, 7 (1961).
[541] HELFRICH, W. and MARK, P., *Z. Phys.*, **166**, 370 (1962).
[542] ADOLPH, J., BALDINGER, E. and GRÄNACHER, I., *Phys. Letters*, **8**, 224 (1964).
[543] GÖRLICH, P., *Advances in Electronics and Electron Physics*, Vol. XIV (C), p. 37, Academic Press (New York) 1961.
[544] ROSE, A., *RCA Rev.*, **12**, 362 (1951); *Photoconductivity Conference* (*Atlantic City*) *1954*, p. 3, J. Wiley & Sons (New York) 1956.
[545] HEILAND, G. and MOLLWO, E., *Landolt-Börnstein*, Vol. II, p. 365, Springer (Berlin-Göttingen-Heidelberg) 1959.
[546] PUTZEIKO, E., *Dokl. Akad. Nauk SSSR*, **132**, 1299 (1960).
[547] GOFF, G. B. F., Brit. P. 804,911, Nov. 26, 1958.
[548] WAITE, T. R., *Phys. Rev.*, **107**, 463 (1957).
[549] MEIER, H., *Z. phys. Chem.*, **208**, 340 (1958).
[550] HEIJNE, L., *Philips tech. Rdsch.*, **25**, 233 (1963/1964).
[551] HELFRICH, W., RIEHL, N. and THOMA, P., *Phys. Letters*, **10**, 31 (1964).
[552] BRYANT, F. J., BREE, A., FIELDING, P. E. and SCHNEIDER, W. G., *Disc. Faraday Soc.*, **31**, 48 (1959).
[553] GROSSWEINER, L. I., *J. appl. Phys.*, **24**, 1306 (1953).
[554] KOMMANDEUR, J. and SCHNEIDER, W. G., *J. chem. Phys.*, **28**, 590 (1958).
[555] GARLICK, G. F. J. and GIBSON, A. F., *Proc. roy. Soc.*, Ser. A. **188**, 485 (1947).
[556] MEIER, H., *Z. wiss. Photogr.*, **51**, 208 (1957).
[557] KALLMANN, H. and ROSENBERG, B., *Phys. Rev.*, **97**, 1596 (1955).
[558] NELSON, R. C., *J. chem. Phys.*, **22**, 885 (1954).
[559] GOBRECHT, H., HAHN, D. and KÖSEL, H. J., *Z. Phys.*, **136**, 285 (1953).
[560] COOK, J. R., *Proc. phys. Soc.*, Sect. B. **68**, 148 (1955).
[561] BREE, A., REUCROFT, P. J. and SCHNEIDER, W. G., *Symposium on Electrical Conductivity in Organic Solids*, p. 113, Interscience Pub. (New York) 1961.
[562] LOWER, S. K. and EL-SAYED, M. A., *Chem. Rev.*, **66**, 199 (1966).
[562a] KOSONOCKY, W. F., HARRISON, S. E. and STANDER, R., *J. chem. Phys.*, **43**, 831 (1965).
[563] MEIER, H., *Angew. Chem.*, **69**, 730 (1957).
[564] WAGNER, C. and HAUFFE, K., *Z. Elektrochem.*, **44**, 172 (1938).
[565] HAUFFE, K., GLANG, R. and ENGEL, H. J., *J. phys. Chem.*, **201**, 221 (1952).
[566] SCHWAB, G. M. and BLOCK, J., *Z. phys. Chem.*, (N.S.), **1**, 42 (1954).
[567] HAUFFE, K., *Reaktionen in und an festen Stoffen* (*Reactions in and on Solids*), Springer (Berlin) 1955.
[568] SUHRMANN, R., *Z. Elektrochem., Ber. Bunsenges. phys. Chem.*, **60**, 804 (1956).
[569] RIENÄCKER, R. and HANSEN, N., *Z. Elektrochem., Ber. Bunsenges. phys. Chem.*, **60**, 887 (1956).
[570] HECKELSBERG, L. F., BAILEY, G. and CLARK, H., *J. Amer. chem. Soc.*, **77**, 1373 (1955).
[571] CHYNOWETH, A. G., *J. chem. Phys.*, **22**, 1029 (1954).
[572] ELEY, D. D. and INOKUCHI, H., *Z. Elektrochem., Ber. Bunsenges. phys. Chem.*, **63**, 29 (1959).
[573] LYONS, L. E. and MORRIS, G. C., *J. chem. Soc.*, 3648 (1957).
[574] LIPSETT, F. R., COMPTON, D. M. and WADDINGTON, T. C., *J. chem. Phys.*, **26**, 1444 (1957).
[575] SCHNEIDER, W. G. and WADDINGTON, T. C., *J. chem. Phys.*, **25**, 358 (1956).
[576] COMPTON, D. M. and WADDINGTON, T. C., *J. chem. Phys.*, **25**, 1075 (1956).
[577] KAUFHOLD, J. and HAUFFE, K., *Z. Elektrochem., Ber. Bunsenges. phys. Chem.*, **69**, 168 (1965).
[578] EPSTEIN, A. and WILDI, B., *Symposium on Electrical Conductivity in Organic Solids*, p. 377, Interscience Pub. (New York) 1961.
[579] DELACOTE, G. and SCHOTT, M., *Phys. stat. sol.*, **14**, 16 and 60 (1962).
[580] HAMANN, C. and STORBECK, J., *Naturwissenschaften*, **50**, 327 (1963).
[581] PETRUZELLA, N. and NELSON, R. C., *J. chem. Phys.*, **37**, 3010 (1961).
[582] LABES, M., SEHR, R. and BOSE, M.,

Proceedings of the International Conference on Semiconductor Physics, Prague 1960, Academic Press (New York) 1961.

[583] MOSER, F. H. and THOMAS, A. L., *Phthalocyanine Compounds*, p. 75, Reinhold (New York) 1963.

[584] PETRUZELLA, N. and NELSON, R. C., *J. chem. Phys.*, **42**, 3922 (1965).

[585] MEIER, H. and ALBRECHT, W., *Ber. Bunsenges. phys. Chem.*, **69**, 917 (1965).

[586] SPENKE, E., *Elektronische Halbleiter (Electronic Semiconductors)*, Springer (Berlin) 1956.

[587] VARTANYAN, A. T. and ROSENSHTEIN, L. D., *Dokl. Akad. Nauk. SSSR*, **131**, 279 (1960).

[588] DMITRIEVSKII, O. D., ERMOLAEV, V. L. and TERENIN, A. N., *Dokl. Akad. Nauk SSSR*, **114**, 751 (1957).

[589] CHO, B. Y., NELSON, R. C. and BROWN, L. C., *J. chem. Phys.*, **39**, 499 (1963).

[590] GLARUM, S. H., *J. phys. Chem. Solids*, **24**, 1577 (1963).

[591] MURRELL, J. N., *Mol. Phys.*, **4**, 205 (1961).

[592] ROBERTSON, J. M. and WOODWARD, Y., *J. chem. Soc.*, 36 (1940).

[593] NELSON, R. C., *J. chem. Phys.*, **30**, 406 (1959).

[594] ROSENBERG, B., *J. chem. Phys.*, **29**, 1108 (1958); **34**, 63 (1961).

[595] TERENIN, A., *Radiotekh. Elektron.*, **1**, 1127 (1956).

[596] KLEINERMANN, M. Y. and MCGLYUN, S. P., *Organic Semiconductors*, edited by J. J. BROPHY and J. W. BUTTREY, p. 108, Macmillan (New York) 1962.

[596a] DRESNER, J., *Phys. Rev.*, **143**, 558 (1966).

[597] KEPLER, R. G., *Organic Semiconductors*, edited by J. J. BROPHY and J. W. BUTTREY, p. 1, Macmillan (New York) 1962; *Phys. Rev.*, **119**, 1226 (1960).

[598] LE BLANC, O. H., *Organic Semiconductors*, edited by J. J. BROPHY and J. W. BUTTREY, p. 21, Macmillan (New York) 1962; *J. chem. Phys.*, **33**, 626 (1960).

[598a] MEIER, H. and ALBRECHT, W., *Umschau*, **67**, 702 (1967).

[598b] PETHIG, R. and MORGAN, K., *Nature, Lond.*, **214**, 266 (1967).

[599] CALVIN, M., *Naturwissenschaften*, **43**, 387 (1956).

[600] WATSON, J. D. and CRICK, F. H. C., *Nature, Lond.*, **171**, 964 (1953).

[601] DREHFAHL, G. and HENKEL, H. J., *Naturwissenschaften*, **42**, 624 (1955); *Z. phys. Chem.*, **206**, 93 (1956).

[602] KRONICK, P. L. and LABES, M. L., *J. chem. Phys.*, **35**, 2016 (1961).

[603] RIEHL, N., BECKER, G. and BÄSSLER, H., *Phys. stat. sol.*, **15**, 339 (1966).

[604] WEIGL, J. W., *J. chem. Phys.*, **24**, 364 (1956).

[605] KEPLER, R. G., BIERSTEDT, P. E. and MERRIFIELD, R. E., *Phys. Rev. Letters*, **5**, 503 (1960).

[606] POHL, H. A. and ENGELHARDT, E. H., *Z. phys. Chem.*, **66**, 2085 (1962).

[607] BERLIN, A. A. and MATVEEVA, N. G., *Dokl. Akad. Nauk SSSR*, **140**, 368 (1961).

[608] SHAROYAN, E. G., TIKHOMIROVA, N. N. and BLYUMENFEL'D, L. A., *Zh. Strukt. Khim.*, **5**, 697 (1964).

[609] LYONS, L. E., *J. chem. Soc.*, 5001 (1957); *Austral. J. Chem.*, **10**, 365 (1957).

VI. MODERN VIEWS ON THE MECHANISM OF SPECTRAL SENSITIZATION

6.1 *Principles of the Theories*

Although, since the discovery of the spectral sensitization of the photographic process, a number of theories have been developed to explain the mechanism of the transfer of energy, this mechanism has remained an unsolved problem practically to the present time.

The discussions originally centered round the formation of a light-sensitive compound of the dye and the silver halide which was said to decompose on exposure in the absorption band of the dye into an atom of silver and a dye residue.[212,610] Various dye-silver compounds which decompose in the said manner are undoubtedly known. However, the finding established by *Eggert, Noddack et al.* to the effect that a dye molecule is capable of forming 100 and more silver atoms before fatigue of the layer sets in through oxidation or other destruction of the dye, provides an unequivocal argument against this chemical theory.

Again, the assumption that fluorescence radiation of short wavelength is responsible for sensitization[611] is not correct, the more so since a (short wavelength) antistokes' radiation is known only in a few cases.

A quantum summation by the dye, followed by the giving up of double quanta,[612] is contradictory to the photographic effect being dependent on the intensity of the light.[42]

At the present time the following two hypotheses are still the main ones discussed. Firstly, the transfer of energy as such in the nature of a resonance process, and secondly the possibility of a direct transfer of electrons from the dye to the silver halide. Common to both these theories is the fact that at the end of every elementary reaction an Ag atom appears and therefore an Ag$^+$ ion must somehow have received an electron. According to the first-named theory, this electron originates from a bromide ion, whilst the last-named electronic theory assumes the transfer of electrons from the dye to the silver halide.

As is shown by the mechanisms and relations discussed in the previous chapter, the mechanism of spectral sensitization cannot be restricted to the question of the formation of electrons in the sensitized substratum. The mechanism must explain how the radiation energy absorbed by the dye forms electrons or positive holes in the light-sensitive system in accordance with the mechanism of conduction in the latter. The formation of these electronic charge carriers indeed ultimately represents the primary step, which can be observed directly by means of photoconduction measurements, and which is then followed immediately by chemical reactions in light-sensitive systems.

6.1.1 THE RESONANCE THEORY. This theory, which has been discussed above all by *Terenin, Akimov* and *Putzeiko*,[404,408,416] regards spectral sensitization as a resonance process in which the energy absorbed by the dye is transmitted to the halide ions with vacant sites and other defects occupied by electrons, and by this means these sites acquire the power of giving up electrons to the silver halide conduction band. According to this theory the primary process in sensitization would therefore be conceived as a radiationless process of energy transfer which is brought about by the remote coupling of the electronic oscillators in the organic dye molecules with the electron oscillators on the surface of the semiconductor. Here the quantum yield of the energy transfer would depend on the strength of the coupling of these oscillators.

The process which is shown diagrammatically in Fig. 37 can be formulated in the following simplified form:

1. $F + h\nu = F^*$
2. $F^* + Br^- = F + Br + \ominus$
3. $Ag^+ + \ominus = Ag$

Fig. 37. Diagrammatic representation of the resonance mechanism of spectral sensitization.

It can be seen that if this mechanism is valid the charge on the dye will not undergo any change, since the displacement in energy only takes place inside the dye-bromide unit.[613] The sole task of the sensitizer is to increase the small probability of a transfer of electrons in the bromide ions containing vacant sites, which share responsibility for the weak red-sensitivity.[310,614]

6.1.1.1 *Arguments in Support of the Resonance Mechanism.*

6.1.1.1.1 Defects. The fact that in a silver halide it is necessary to reckon with the presence of halide ions containing vacant sites and other defects from which electrons can be excited into the conduction band of the silver halide, indicates that a resonance mechanism is possible, and this is supported by the long wavelength "tail" absorption.

Furthermore, *Terenin et al.*[251,408,418] regard the promotion of the sensitizing action exerted by adsorbed electronegative gases (Br_2, I_2) as a further argument in support of the said mechanism. Moreover, the authors assume that during the absorption of light by the dye, an electron passes by means of resonance from the silver halide valency band to the acceptor molecule from whence, by means of further excitation, it then gets into the AgBr-conduction band.

6.1.1.1.2 Radiationless Transfer. The fact that in various systems energy can be transmitted without radiation from one excited molecule to another might nowadays be taken as proved. The evidence given by a number of studies of fluorescence is in conformity with a process of this kind.

Such studies include, for instance, those of the sensitized fluorescence of molecule crystals which have been doped with aromatic hydrocarbons,[615-618] observations of the quenching of the fluorescence of fluorescing substances dissolved in viscous media[225,618,619] or measurements of the quenching of fluorescence in crystals of anthracene or naphthalene on embedding traces of foreign molecules in them.[620-623]

The experiments of *Kuhn et al.* [624-626] can be regarded as giving direct proof that radiationless transfer of energy is possible. In these experiments one layer of dye excites fluorescence in a second absorbing layer although both layers are separated by thin layers of barium stearate. When using pairs of dyes which fulfil the prerequisite for radiationless transfer, namely that the maximum fluorescence of the sensitizer occurs in approximately the same position as the maximum absorption of the acceptor which takes up the energy,[225,619] transfer of energy of up to a maximum distance of $d_0 = 300$ Å was moreover observed. In these experiments the radiationless transfer of energy could be recognized not only from the occurrence of fluorescence in the acceptor, but also from the decomposition of the latter—which is due solely to the absorption of radiation by the sensitizer—or in the case of a non-fluorescing acceptor, also directly from the quenching of the fluorescence of the sensitizer.[626]

The experimental values obtained by *Kuhn* and his collaborators for the distances over which energy is transferred without radiation from an excited molecule S to an acceptor A are remarkably in accordance with those to be expected from *Förster's* theory.[225,619,627-629,629a] Since *Förster's* theory attributes the migration of the excitation energy to a radiationless, direct electrodynamic interchange (resonance) between the primarily excited molecules (oscillators S) and the neighbouring molecules (oscillators A), the possibility of a transfer of energy without radiation from a photographic sensitizer to the surface of a semiconductor must therefore, on the basis of *Kuhn*'s proof, be included in the discussion.

6.1.1.1.3 Position of the Energy Levels. A further reason for accepting the resonance mechanism goes back to the calculations and measurements made by *Coulson, Akimov et al.*,[251,408,630,630a] according to whom the excitation level in a dye should lie about 4 to 6 eV below the ionization limit, that is under the AgBr-conduction band which is at a distance of about 3·5 eV from the ionization limit. The transfer of electrons, which presupposes that the conduction band of AgBr is situated below the excitation band of the dye is therefore improbable.

Even the two-electron theory discussed by *Mitchell*[631] was unable to provide any explanation for the transfer of electrons because it was too hypothetical.

This theory states that, on absorbing light, the dye molecules become negatively charged by taking up electrons from Br⁻ ions containing vacant sites, and this raises the energy levels to such an extent that a transfer of electrons from the dye to the acceptors in the AgBr and from the latter by thermal excitation into the conduction band of the AgBr appears possible.

6.1.1.2 *Arguments against the Resonance Mechanism.*

6.1.1.2.1 The Influence of Defects. The presence of vacant sites which result from the building in of foreign atoms, thermal transfers inside the lattice (e.g. Frenkel-vacant sites) or from various imperfections in the structure of the crystals (especially in polycrystalline powders and layers) have certainly to be reckoned with in silver halide emulsions: cf. moreover also.[42,338,454,632,633-636]

This is not necessarily contradictory to the hypotheses which assume the presence of bromide ions which are rich in energy (electron donors) in crystal dislocations, on the surface of the grains and in other places, and in which moreover the influence of adsorbed bromine atoms (electron acceptors) is discussed.[403]

The following objections can, however, be raised to the interpretation of the action of these defects given by the resonance theory:

1. The probability of a radiationless transfer of energy and therefore of the sensitization action would have to be increased owing to the installation of additional defects which act as electron donors in accordance with equation 58 (see the next section). The observation made by *Carroll et al.*[264,265] that the sensitization reaction was independent of the chemical sensitization (see section 1.2.8 of Chap. III) shows, however, that this is not the case. Even the influence of adsorbed gases and vapours (such as bromine) as established by *Terenin et al.* does not offer a solution to this contradiction.

2. The action of adsorbed bromine acceptors would, according to the resonance theory, consist firstly in an electron from the silver halide valency band being excited into the acceptor term by the radiationless transfer of energy. The initiation of the primary photographic process by the transfer of electrons from a negatively charged "acceptor" into the conduction band of AgBr could not take place until a second light-quantum had been absorbed by the dye, or by means of thermal energy. Therefore in certain circumstances the spectral sensitization effect would have to represent a 2-quanta process.[403] Various observations have shown, however, that a 2-quanta process is ruled out from the outset, firstly because in some cases experimental values of the order of magnitude of 1 have been obtained for the quantum yield in spectral sensitization,[36,637] and secondly because the reciprocity law failures in the region of absorption of the dye correspond with those which occur in the spectral region of the normal absorption of AgBr.[638]

Any kind of thermal excitation would undoubtedly reveal itself by producing a temperature effect on the spectral sensitization reaction and this effect would differ greatly from that in the normal absorption region. The measurements made by *Frieser, Graf* and *Eschrich*[163a,190] indicate the existence of this temperature relation, whereas the results of other experiments are contradictory to this finding.[404,410]

6.1.1.2.2 The Problem of the Radiationless Transfer of Energy. When discussing the resonance theory of spectral sensitization it may be assumed that the same relations can be used for calculating the probability of a radiationless transfer of energy from the dye layer to the bromide ions containing vacant sites as those which were derived by *Förster, Kuhn et al.*[625,629] for the radiationless transfer of energy in general.

It can be seen from equation 58 that the probability $N_{D \to La}dt$ of an excited dye molecule D giving up its excitation energy to the lattice defects La(Br^- etc.) during a time interval of dt is greatly dependent on the distance d between the dye and the lattice defect.

$$N_{D \to La}dt = \left(\sigma \cdot \gamma \cdot C \frac{1}{d^4}\right) dt \tag{58}$$

In this equation σ is the number of lattice defects which take up the energy per unit area, $\gamma = \pi/3$ (when the directions of the D and La oscillators respectively are statistically distributed in space) and

$$C = \frac{9 \cdot \ln 10 \cdot c^4}{128 \cdot \pi^5 \cdot N_L \cdot \tau_s^* \cdot n^4} \cdot \int_0^\infty f_D(\nu)\varepsilon_{La}(\nu) \frac{d\nu}{\nu^4} \tag{59}$$

where c is the velocity of light, N_L, the *Loschmidt* number, τ_s^*, the natural mean decay period of the fluorescence of D, n, the refractive index of the intermediate medium, f_D, the distribution function of the quantum spectrum of the fluorescence of D as standardized in accordance with $\int_0^\infty f_D d\nu = 1$, ε_{La} is the molar decadic extinction coefficient of La, and ν is the frequency of light.

Since the probability of the excitation energy of a sensitizing dye being given up by means other than radiationless transfer during a time interval of dt—e.g. by natural fluorescence—is given by

$$N_{D \to La}dt = \left(\frac{1}{\tau_s}\right) dt \tag{60}$$

(where τ_s is the mean decay period of the fluorescence of the non-adsorbed dye), then the thickness of the layer in the case in which a radiationless transfer and intrinsic fluorescence are equally probable ($d = d_0$) will be given by:

$$d_0 = \sqrt[4]{\sigma \cdot \gamma \cdot C \cdot \tau_s} \tag{61}$$

Kuhn, et al.[624–626] carried out experiments on suitable systems which were built up of two layers of dye capable of fluorescing placed between multimolecular layers of Ba stearate, and demonstrated that at a distance of $d = 300$ Å the fluorescence band of the sensitizer was about half its size at greater distances (i.e. $d_0 = 300$ Å).

The fluorescence of the sensitizer does not however become completely quenched in the way that is observed with sensitized inorganic systems (see Chap. III, 1.2.8), until the distance d becomes equal to 50 Å, whereas, when the distance d exceeds 500 Å there is absolutely no indication of any radiationless transfer of energy.

There is no doubt that the acceptance of these experimental values obtained with model systems in which the lattice defect-oscillators are replaced by dye-oscillators, appears somewhat problematical. For instance, there is a complete absence of any data regarding σ and ε_{La} (and therefore C) of the lattice defect-oscillators. The investigations do, however, prove that a quantitative radiationless transfer of energy is possible up to distances of 50 Å and, with a smaller yield, up to about 300–500 Å.

How can this finding be brought into harmony with the properties of spectrally sensitized photographic emulsion layers?

1. Since the distance at which a radiationless transfer of energy cannot possibly exceed $d_{\text{limit}} = 5 \cdot 10^{-6}$ cm, measurements which show that spectral sensitization can still take place with layers of sensitizer of $1 \cdot 10^{-4}$ cm in thickness (see Chap. III.1.2.4) cannot be explained by means of the resonance hypothesis. For instance it does not explain why the efficiency in sensitization increases with increase in the thickness of the dye coating up to $6 \cdot 10^{-5}$ cm[638a] in the case of selenium.

2. The observation that a layer of dye can still be excited to fluorescence by the radiationless transfer of the energy absorbed by a sensitizer, even when it is separated by layers of over 50 Å in thickness, would lead one to expect that the sensitizing action was independent of the manner in which the sensitizer was deposited. The effects of sensitized layers discussed in chapters II and III indicate however that it is precisely the contact between the dye and the silver halide grain which is the essential factor for producing the sensitizing effect. Every influence, whether it be conditioned by the constitution of the dye or by additional, more strongly adsorbed molecules, either weakens or annuls the sensitizing power of a dye. Since, according to the theory of the radiationless transfer of energy, the sensitizer molecules are largely independent of their orientation, the excitation energy, at least over a distance of 50 Å, should be passed on to the statistically distributed lattice defects, it must be concluded from the behaviour of sensitized photographic emulsion layers that the energy is transferred by another mechanism during the reaction of photographic sensitization.

3. In contradiction to the resonance mechanism are also the investigations of *Kuhn, Nelson, Schäfer* and *Drexhage*,[639] in which the complete destruction of spectral sensitization on interposing a monomolecular layer of barium stearate

between the dye and the photoconductor (CdS) was revealed. However, in silver bromide crystals which were covered by a cyanine dye-cadmium arachidate layer system both sensitization and quenching of fluorescence could be observed for separation distances of the order of 50 Å.[638b] But it must be noted that it is impossible to decide exactly from experiments of this kind whether the mechanism is one of resonance or of electron transfer. The uncertainty arises because even when spectral sensitization takes place when there is an insulating layer between the sensitizer and the semiconductor, the possibility of a charge transfer taking place across the layer cannot be ruled out. The observations of barrier layer effects in systems such as Cu/insulating organic material (polymer compounds, etc.)/Cu-O proves that electrons or holes must be able to cross insulating layers in some cases.[639a,639b]

4. Even if any interpretation could be put on the decrease in the sensitizing action with increase in layer thickness in accordance with equation 58, and from the theory of the radiationless transfer of energy, the problem of the conversion of a sensitizing effect into one of desensitization—which is not due to any filter action—in the case of thick layers, still remains unsolved. The theory of the radiationless transfer of energy is also unable to provide an answer to the desensitization effect produced by certain dyes—see Chaps. II and III.

5. Moreover, a characteristic feature of any transfer of physical energy—and not only of the singlet-singlet resonance-energy transfer corresponding to equation 59, but also of the triplet-triplet short range exchange-energy transfer in which the donor and the acceptor are sufficiently close to one another for their electron clouds to overlap and which is very important for the sensitization of organic photochemical reactions in solution (see e.g. Hammond et al.[639c-639f])—that among other things the energy gap between ground state and excited state of the donor is greater than the corresponding energy gap of the acceptor. Therefore, the behaviour of donor dyes with similar absorption bands should be analogous and the differences in the mode of charge injection which are observed with similar dyes in many cases cannot be explained with the energy transfer mechanism. It cannot explain, for instance, why the spectrally sensitized photocurrents in selenium correspond to the injection of electrons into the body of the photoconductor when Crystal violet and Pinacyanol are used as sensitizers, but correspond to the injection of holes when Methylene blue, Phenosafranine and nitrosobenzene are used as sensitizers.[638a]

6.1.1.2.3 The Position of the Energy Levels. The position of the excitation level in a dye which was calculated by *Coulson et al.* as $7\,\text{eV} - h\nu_{\max}$ and which appears to exclude the transfer of electrons from the dye to the silver halide, does not correspond with the actual energy relations in the dye, as the later statements show. In this connection it is pointed out that according to *Scheibe's* rule,[640-642] the first excitation single level in dyes can be assumed to be about 3·4 eV below the ionization limit.

There is therefore no necessity for the resonance theory[53,269] to be discussed on the grounds of the energy relations.

6.1.2 THE ELECTRONIC THEORY OF SENSITIZATION. In contrast to the resonance hypothesis in which it is assumed that the electron which is responsible for the reduction of an Ag^+ ion comes from the bromides which are favourably situated as regards energy in the grain, the electronic theory assumes that electrons are transferred from the dye to the AgBr; cf. Fig. 38.

Whereas—apart from long exposures during which the bromine formed can have a deleterious effect—the resonance process manages to dispense with any special stage of regeneration, the electronic mechanism assumes, on the basis of the repeated reactivity of the dye molecules, that the charge on the dye is compensated by the return of electrons from the AgBr. The discussion therefore centres round the following scheme.[269,643,644]

1. $F + h\nu = F^+ + \ominus$
2. $Ag^+ + \ominus = Ag$
3. $F^+ + Br^- = F + Br$

The electronic mechanism of sensitization was first postulated by *Gurney* and *Mott*,[309,312] but in later years it was regarded by these and other authors as being improbable.[113,266,310,613] Firstly, the transfer of electrons from the dye

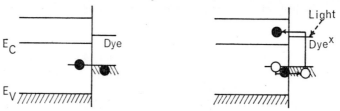

Fig. 38. Diagrammatic representation of the electron transfer mechanism of spectral sensitization.

to the silver halide appeared to be unlikely because the ionization energy of the sensitizer was too high. Secondly, for the hypothesis to be valid, it was necessary to assume the formation and migration of photoelectrons in a pure sensitizing dye, an effect for which a proof was likewise not forthcoming at the moment when the electron transfer hypothesis was being rejected.[36,645]

On the basis of more recent experimental findings these counter-arguments are however no longer tenable, so that the use of the mechanism proposed by *Gurney* and *Mott* as a hypothesis for the transfer of energy in sensitized photographic emulsion layers appears to be open to discussion. The following part-reactions can, moreover, be proved to take place in accordance with the original electron transfer reaction:

1. The formation of a photoelectron (and of a hole) in a dye by means of a $\pi-\pi^*$ excitation, on exposing in the absorption band of the dye.

2. The migration of the excitation energy over several dye molecules, a process which is of importance precisely when polymerically adsorbed dyes are present.
3. The transfer of a photoelectron from the excitation level and from the conduction band, respectively, of the dye into the conduction band of AgBr.
4. The return of an electron from the AgBr to the dye.

It can be seen that with this mechanism the sensitizer does not itself intervene in the course taken by the actual photographic primary process. The dye alone has the task of placing an electron which has been brought by its own absorption into the conduction band of AgBr in the case of an unsensitized emulsion layer.[216,280,634,646,647,648] The extent to which the formation of a hole in the valency band of AgBr immediately after the bromide ions with vacant sites in the AgBr have given up electrons—which ions may also take part in the process when the dye is regenerated—is still a debatable question. In this connection it should be mentioned that according to *Mitchell*'s theory of the formation of the latent image, the production, not only of photoelectrons, but also of holes, is considered to be important for the photographic primary process, since the latter form the interstitial silver ions Ag_0^+.

Any discussion of this mechanism must not overlook the fact that *p*-conducting semiconductors can be sensitized directly has not been explained. The mechanism provides an answer to the question of the formation of electrons in the conduction band of silver halides and semiconductors, but it neglects the fact that above all the concentration of the positive holes in sensitizable *p*-conductors increases on exposing the sensitizer. Moreover, the electronic theory of sensitization in its original simple form leaves the question of desensitization or supersensitization unanswered, and also gives no indication as to why, for example, photographic desensitizers are able to sensitize photoelectric effects in various photoconductors.

This means that the mechanism of the transfer of electrons will have to be supplemented if it is to explain the results obtained in studies of photographic sensitization and desensitization and of the closely related sensitization of photoelectric effects in *p*- and *n*-conductors.

6.1.3 COMMENT ON THE QUESTION: IS THE REACTION ONE OF RESONANCE OR OF ELECTRON TRANSFER? Practically the only difference between the two mechanisms of sensitization at present under discussion is that an electron which is brought into the conduction band of a silver halide originates either from the dye or from the silver halide grain. Even if the electrons in the dye cannot be distinguished experimentally from those in the semiconductor any desire to lay aside the question of the mechanism of the transfer of electrons as being unanswerable must however be resisted.

An explanation is necessary in order to meet the practical problems arising in sensitization, notably in deciding whether or not various dyes can be used for spectral sensitization. For example, when selecting dyes for the sensitization of semiconductor materials of interest in electrophotography, those which would be considered suitable if the resonance hypothesis were the correct one would not be regarded as such if the electron transfer reaction were valid, which latter stipulates for instance close contact between the semiconductor and the dye for sensitization to take place. A theoretical comprehension of the mechanism of supersensitization which is linked with sensitization could lead to this reaction being made use of to a larger extent than previously for raising the yield in spectrally sensitized systems—e.g. in the infra-red region of the spectrum. The ability to predict the sensitizing or desensitizing power of a dye without having to make long empirical tests likewise demands a knowledge of the mechanism of sensitization.

6.2 The Electron Transfer Theory of Spectral Sensitization

6.2.1 THE PROBABILITY OF THE TRANSFER OF ELECTRONS BETWEEN A DYE AND A SEMICONDUCTOR. When discussing the electronic theory of sensitization it is assumed that by so doing the question of the extent to which electrons can be transferred from a dye to AgBr even in the presence of a potential wall between the AgBr and the dye will be elucidated.[613] An appraisal in accordance with the theory of the tunnel effect which gives a general description of the

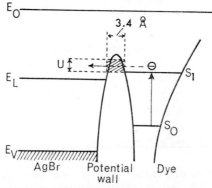

Fig. 39. Diagrammatic representation of the potential wall between a silver halide and a dye.

passage of electrons through threshold potentials—cf. [53,252,643,650]—gives the following result:

The rate constant of the tunnel transfer k_T: The rate constant k_T of the tunnel transfer of an electron through a potential wall between a dye and AgBr—simplified according to Fig. 39 by being assumed to be rectangular

—is given by the product of the permeability coefficient D and the number of impacts z:

$$k_T = D \cdot z \qquad (62)$$

Moreover, the following equation holds for the permeability coefficient D[252,650]:

$$D = D_0 \cdot e^{-\frac{2}{h}\sqrt{2m(U-E)} \cdot l} \qquad (63)$$

In equation 63, D_0 is a proportionality factor of the order of magnitude of 1 according to[650]; $h = 6{\cdot}626 \cdot 10^{-27}$ (erg . sec); $m_\ominus = 0{\cdot}9107 \cdot 10^{-27}$ g (the mass of the electrons); U is the estimated threshold energy of the potential wall (≈ 2 eV); E is the thermal energy of the electron ($\approx 0{\cdot}1$ [eV]) according to[252]; l is the length of the potential wall which corresponds with the distance between the adsorbed dye and the silver halide, and which, according to *Scheibe* and *Dörr*[252] is about $3{\cdot}4$ Å.

With the values given, the numerical value of the permeability coefficient D is of the order of between 10^{-3} and 10^{-2}.

The number of impacts z of an electron excited in the dye term S_1 is given by equation 64:

$$z = \frac{v}{s} \qquad (64)$$

where v is the formal, spatial velocity of an electron which is situated in the excitation level, and s is the length of the conjugated chain in the dye molecule (length of the electron path). Since the electron impulse is given by

$$\Delta p = m \cdot v \qquad (65)$$

and moreover

$$\Delta p \cdot s = \frac{h}{4\pi} \qquad (66)$$

the number of impacts z can be estimated from the relation 67:

$$z \approx \frac{h}{4\pi \cdot m \cdot s^2} \qquad (67)$$

according to which z is of the order of magnitude of $5 \cdot 10^{13}$ (sec^{-1}), so that a value of $k_T \approx 10^{11}$ for the rate constant is debatable.

The rate constant of the return of an excited electron (k_R): In the case of dyes the probability of an electron returning to the ground state can be derived from the rate constants of the fluorescence emission:

$$k_R = \frac{1}{\tau_F} \qquad (68)$$

Since the lifetime of the excited electrons is $\tau_F \approx 10^{-8}$ (sec), the rate constant k_R can be assumed to have a value of $k_R \approx 10^8$ (sec^{-1}).

Comparison of k_T with k_R: A comparison of the rate constants k_T and k_R ($k_T/k_R \approx 10^3$) shows that the probability of the transfer of an excited electron to the conduction band of silver halide is greater by a factor of 10^3 than that of the return of the electron to the ground state of the dye. That is to say, even assuming a relatively high potential wall, the possibility of electrons being transferred from the dye to the silver halide is a debatable point.

Since the height of the potential wall which is assumed in the example to be ($U \approx 2$ eV) hardly corresponds with the conditions obtaining in the AgBr/dye system when the sensitizing dyes are strongly adsorbed—in this case the influence of a potential wall can possibly be neglected completely[53,269]—the probability of electrons being transferred might be still greater than that of the return of the excited electron, either with or without the emission of radiation. In the case of dye molecules which for structural or other reasons (the influence of gelatin, etc.) are not in close contact with the silver halide grain, D and therefore k_T will however decrease greatly owing to an increase in l in accordance with equation 63, so that the probability of the transfer of electrons and therefore of sensitization—in opposition to the theory of a radiationless transfer—must decrease perceptibly when the taking up of dye is unfavourable.

It should also be noted that the estimation of the probability of electron transfer explains the observed decrease in the fluorescence of dyes on adsorption to silver halides and other photosemiconductors. The electron transfer theory of sensitization is therefore in harmony with the observed quenching of fluorescence.

6.2.2 THE RELATION BETWEEN THE CONDUCTIVITY OF SENSITIZING DYES AND SPECTRAL SENSITIZATION. It has already been mentioned that it has been found impossible to establish a direct relation between the order of magnitude of the photoconductivity of a dye (see Chap. V) and its sensitizing action. This is readily understandable since, before the sensitizing action can take effect, a number of prerequisites—close contact between the dye and the AgBr, suitable position of the energy levels, etc. (see Chap. III)—which, when measuring the photoconductivity of a dye are of no importance whatever, have to be satisfied.

There does however exist a series of observations which permit a relation to be recognized between the various regularities in the behaviour of spectral sensitization on the one hand and the corresponding effects of the dark- and photo-conductivity of a sensitizing dye on the other hand. This might indicate that the electronic conductivity behaviour of a dye has an effect on the sensitization reaction and that this reaction is therefore based on an electronic mechanism.

6.2.2.1 *The Dark-conductivity of Dyes and Sensitization. Nelson*[651] was the first to report on a relation between dark conductivity and sensitizing action. Taking Pinacyanol as an example, which dye can be prepared by a special technique in a good and in a poorly conducting form,[505] the author was able to establish that the Pinacyanol modification with the lower dark conductivity was a good sensitizer, whereas the modification with the higher (by about a factor of 10^3) dark conductivity either did not sensitize at all or did so only to a very slight extent.

Even if this relation is only of a qualitative nature, it does however show that the conductivity behaviour of unexposed dyes can have an effect on their sensitizing power.

6.2.2.2 *The Relation between Layer Thickness and Sensitizing Power.* Both the quantum yield of the sensitization reaction and the quantum yield of the photoconductivity increase with decrease in the thickness of the layer of dye (see Chap. III, 1.2.4 and Chap. V, 3.5). The rule that a maximum sensitizing action is achievable with monomolecularly adsorbed layers of dye does not therefore stand in contradiction to the photoelectric behaviour of the dye, since the quantum yield of the photoconductance might also reach a maximum in the case of monomolecular layers of dye.

This is indicated not only by the results given in section 3.5, but also by the measurements made by *Kuhn* and his collaborators[626] of the photoconductance of polythienylene [66] built monomolecularly into barium arachidate.

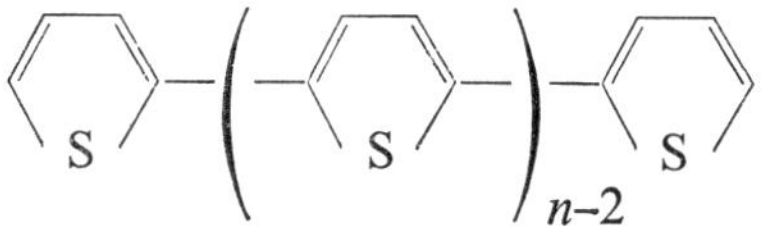

[66]

Since the activation spectrum of barium arachidate/dye sandwich type cells differs from the absorption spectrum of the orientated dye deposit on the arachidate layer, but does conform with the spectrum in solution, it is assumed that a minority of molecules which are not perceptible in the absorption spectrum is already sufficient to produce photoconduction. The argument put forward that the individual dye molecules may have a sensitizing action is in conformity with this observation.[404]

6.2.2.3 *The Relation between Sensitizing Power and Temperature.* The noticeable dependence on temperature of the photographic sensitivity of a sensitized in comparison with an unsensitized AgBr emulsion, and also the analogous effect of temperature on the photoconductivity of TlI can possibly, as shown by the results given in section 3.7 of Chap. V, be brought into relation with the photoconductivity of the sensitizer. Since the photoconductivity of dyes increases with rise in temperature, in the event of the sensitizing action combining with the photoconductance of the dyes, the dependence of the sensitization reaction on temperature might possibly result from a corresponding effect in the dye.

6.2.2.4 *The Relation between Sensitizing Power and Wavelength.* The sensitization reaction and the photoconductivity of dyes show a corresponding dependence on wavelength. Dye aggregates which, in the case of the cyanines, form preferentially on the silver halide surface and to which a marked sensitizing action is ascribed, can moreover also be detected in the activation spectrum as J-bands. This proves that dyes in various forms of aggregation can be photoconductors and sensitizers.

6.2.2.5 *Supersensitization*.

6.2.2.5.1 The Connection between Supersensitization and the Photoconductance of a Dye. The changes which are observed in the dark and photoconductivity of various dyes on adding special organic compounds (*o*-chloranil, etc.) reveal that the supersensitization reaction is based on an interchange of

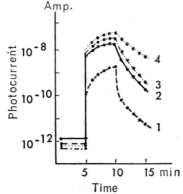

Fig. 40. The influence of doping on the photoconductivity of ZnO. After *Inoue* and *Yamaguchi*.[69] (1) ZnO + Rose Bengals. (2) (1) + 0·01 mol % *p*-chloranil. (3) (1) + 0·01 mol % I_2. (4) (1) + 0·01 mol % phthalic anhydride.

electrons between a supersensitizer and a dye, and this interchange probably influences the conductivity of the dye on the sensitized substratum (AgBr, ZnO, etc.). The supersensitization reaction therefore appears to be particularly suitable for testing the connection between the sensitizing power and the photoconductivity of a dye.

This state of affairs is illustrated in Fig. 40, which represents the increase in the sensitized photoconductivity of ZnO produced by the addition of electron-acceptors (*o*-chloranil, etc.) as measured by *Inoue*,[69] and by way of comparison, the increase observed in the photoconductivity of a dye-sensitizer brought about by the electron-acceptors is given in Fig. 41. (The sensitizers, the Rose Bengals and Merocyanine FX 30 [62] number among the *p*-conductors.)

This suggested also making use of the supersensitization effect as a criterion for deciding the applicability of a sensitization hypothesis, since it may be assumed that that hypothesis which best reproduces the experimental findings is also the one which predicts the increase in the sensitization yield on increasing the concentration of charge carriers in the sensitizer.

This problem will be discussed in greater detail in Chap. VII. In the following the first question to be considered will be the mechanism by which the dark- and photoconductivity of a sensitizing dye is increased by the addition of traces of certain organic compounds (doping effect).

6.2.2.5.2 *Rules Governing the Doping Effect.* The effect of doping has occasionally been tested by measuring the photoconductivity of a dye before and after the addition of a doping agent. The doping agent was added either by sublimation or by spraying a dilute solution of the doping compound on to dye cells. It should be pointed out that this process does not yield any exact

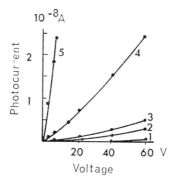

Fig. 41. Merocyanine FX 30 (Formula, see Table 8). Relation between the photocurrent and the voltage and the doping. $1\cdot 8 \cdot 10^3\ \mu W/cm^2$, unfiltered light (4000–8000 Å) surface type cell. (1) Undoped dye. (2) (1) + dibromosuccinic acid (10^{-7} mol/cm²). (2) (1) + tetraiodophthalic anhydride (10^{-7} mol/cm²). (3) (1) + o-chloranil (10^{-7} mol/cm²). (5) (1) + iodine (10^{-7} mol/cm²).

data of the distribution of the doping agent in the dye. The data of the quantity of doping material (mole/cm²) only signify that in each case under the same conditions x moles of the substance are deposited on an electrode screen of 1 cm² in area.

There is therefore no doubt that complete mixing of the dye molecules and the doping agent—in the sense of true doping—cannot be guaranteed in these experiments. However, the experiments do reveal the essential features of the effect of doping, as is confirmed by measurements of high polymers, with which thorough mixing is achieved by the method of manufacture employed:

1. The photoconductivity of merocyanine dyes and phthalocyanines can be raised by a factor of 5, to $5 \cdot 10^4$ by the addition of traces (of the order of 10^{-7}–10^{-8} mole/cm²) of the electron-acceptors compiled in Table 16. Cf. moreover, Table 17.

Like the fluorescein dyes (Rose Bengals) which were tested by *Inoue* in connection with the effect of supersensitization, the dyes of these classes also belong to the *p*-conductors, as has been proved by measurements of the dependence of supersensitization on O_2 and H_2, respectively,[53,487,644] and of the *Hall* effect.[513,578,579]

2. The various electron-acceptors differ greatly from each other in their doping efficiency, as can be seen for instance from the measurements shown

TABLE 16

A SURVEY OF SOME OF THE ELECTRON-ACCEPTORS WHICH ARE SUITABLE FOR DOPING *p*-CONDUCTING DYES

Compound	Formula	Compound	Formula
Iodine		Tetracyanoethylene	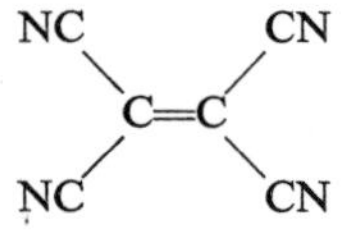 (NC)₂C=C(CN)₂
o-chloranil	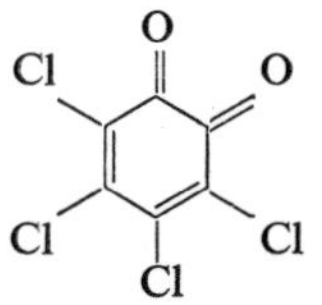	*p*-chloranil	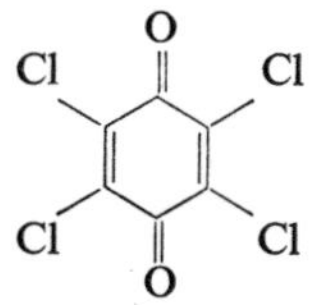
Tetraiodophthalic anhydride	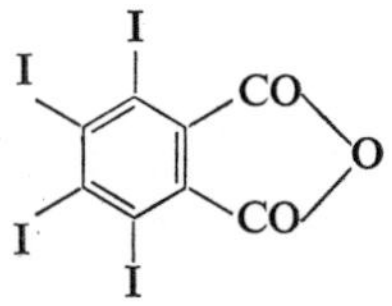	*p*-bromanil	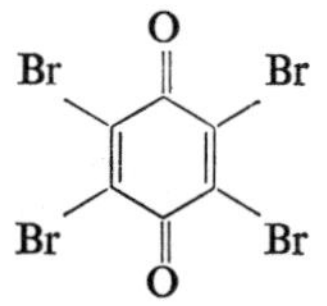
Dibromosuccinic acid	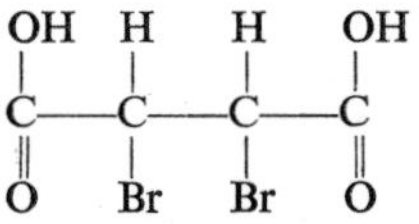	*m*-dinitrobenzene	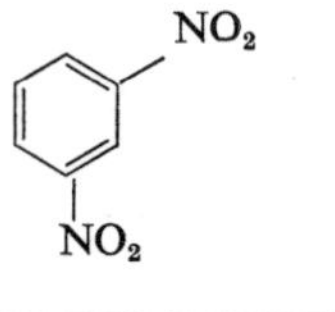

in Fig. 42. Iodine and *p*-chloranil, followed by tetraiodophthalic anhydride, dibromosuccinic acid, tetracyanoethylene, *o*-chloranil, *o*-bromanil and *m*-dinitrobenzene exert the strongest doping action on the merocyanines.

3. No change has yet been observed in the type of conduction.[653] Merocyanines and phthalocyanines remain *p*-conductors even after having been doped.

4. The doping action is dependent on the quantity of the doping agent. As shown for instance by the series of measurements given in Figs. 43 and 44, the photoconductivity increases with the increase in the quantity of the doping agent, and finally approaches a saturation value.

When this saturation limit is exceeded the photoconductivity again decreases but the dark conductivity undergoes a further increase. The specific dark conductivity can moreover reach values of up to $10^{-2}\ \Omega^{-1}\ cm^{-1}$ (undoped: $10^{-10}\ \Omega^{-1}\ cm^{-1}$) with phthalocyanines for example.

TABLE 17

Dye	Formula	Doped with	$F = \dfrac{I_{\text{phot}}\ doped}{I_{\text{phot}}\ undoped}$
Agfa 10	see Table 8	o-chloranil	60
		iodine	1750
		m-dinitrobenzene	7
Merocyanine FX 30	see Table 8	o-chloranil	150
		p-chloranil	6
Merocyanine FX 36	[67]	o-chloranil	50
Merocyanine Rie 54/3	[68]	tetraiodophthalic anhydride	100
		dibromosuccinic acid	100
		Rhodamine B	—
Phthalocyanine		Tetracyanoethylene	120^{652}
			400
		o-chloranil	$5 \cdot 10^{4\ 652}$

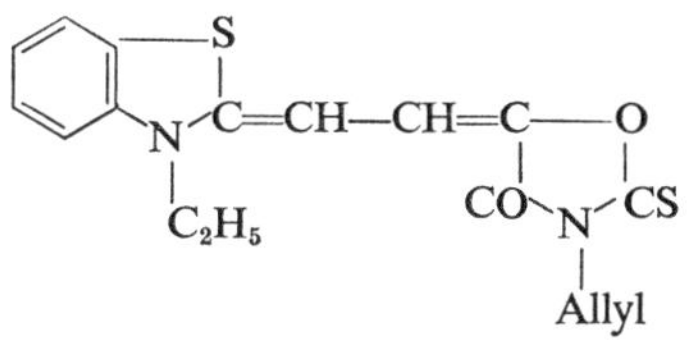

[67]

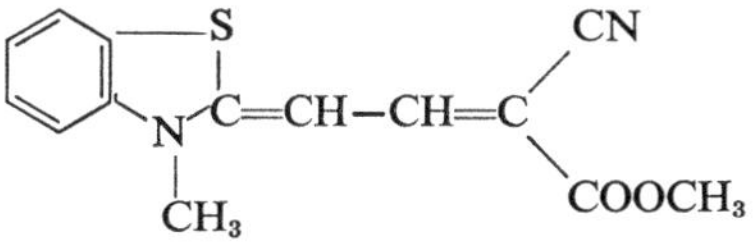

[68]

5. The quantum yield G—see section 3.5 of Chap. V—is raised by several orders of magnitude by the addition of electron-acceptors. For instance, measurements of the saturation current have shown that the quantum yield of Merocyanine FX 30 [62] increases from $G = 10^{-2}$ to $G = 0.6$, i.e. the effect of doping less efficient dye-photoconductors is to turn them into good photoconducting dyes comparable with Malachite green or Pinacyanol (G: 0·3 and 0·5 respectively).

It must be emphasized that values of G of even greater than 1 can be obtained by doping; e.g. when iodine was added to the Merocyanine dye Agfa 10 [59] values of G equal to 2·3 were obtained.

6. The laws governing the photoconductivity of undoped dyes—see section 2.2.3 of this chapter—are not altered by doping. Fig. 48 shows for example

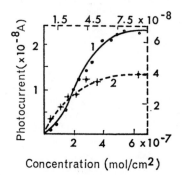

Fig. 42. Merocyanine FX 30 (Formula, see Table 8). Relation between the photocurrent and the concentration of the doping agent. (1) Dye + o-chloranil. $\lambda = 5495$ Å, $157 . 10^{13}$ quanta/sec cm². (2) Dye + iodine. Unfiltered light (4000–8000 Å) $1·8 . 10^3$ μW/cm². Note: The statement mol/cm² signifies that in each case under the same conditions x mols were brought on to an electrode screen cell of 1 cm² in area.

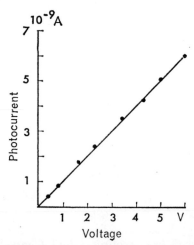

Fig. 43. Merocyanine A10 + iodine (Formula, see Table 8). The photocurrent as a function of the voltage. Surface type cell (distance between the electrodes 0·01 cm). $\lambda = 6000$ Å. $120 . 10^{13}$ quanta/sec cm².

that at low field strengths, Ohm's law also holds for doped dyes: see moreover [653,55].

7. Dyes of the n-type (e.g. Rhodamine B) cannot be doped with electron-acceptor compounds.

8. The effect of doping is not limited to dyes but can in all likelihood be regarded as an effect which is valid for most organic photoconductors. The results obtained here, as well as the measurements made by *Hoegl et al.*[55,605,654–665] with Luvican, multinuclear hydrocarbons (Violanthrene, Perylene), etc., indicate that

> p-conducting organic compounds can be doped with electron-acceptors, and n-conducting organic compounds with electron-donors.

When this rule happens to be combined with supersensitization it can be assumed that

> when inorganic or organic photoconductors (and photographic emulsion layers) are sensitized by dyes of the p-type (n-type) respectively, supersensitization can be achieved above all with electron-acceptor compounds (electron-donor compounds).

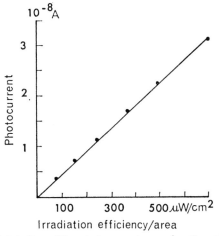

Fig. 44. Merocyanine A10 + iodine. The photocurrent as a function of the irradiation intensity ($\lambda = 6000$ A).

The measurements made by *Inoue et al.*[69,451] are in conformity with the last-named assumption; their universal validity cannot however be established until further experiments, which should above all include actual photographic supersensitizers (see Chap. III), have been carried out.

6.2.2.5.3 Interpretation of the Effect of Doping. Generalities: the effect of doping is in all probability based on a transfer of electrons from the dye, which acts as the donor, to the acceptor molecules.

The transfer must therefore be conceived as a light-induced "charge-transfer-process" (I) using the meaning attached to it by *Mulliken*,[666] and also as the result of the acceptor action of foreign molecules (II) such as that which is also observed in the case of inorganic semiconductors:[667]

I. *CT*-complex between the dye (*D*)—the donor—and the acceptor (*A*):

$$DA \rightleftarrows D^+A^-$$
$$\updownarrow h\nu$$
$$DA \rightleftarrows D^+A^-$$

II. "Trap"-action of the acceptor A (I_2, *o*-chloranil, etc.)

$$D + h\nu \rightarrow D^*$$
$$D^* + A \rightarrow D^+ + A^-$$

As illustrated by the diagram in Fig. 45, the acceptor molecules become negatively charged by these transfers I or II and this raises the concentration

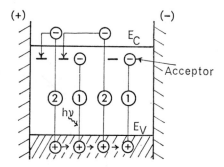

Fig. 45. Diagram in explanation of the doping action. (1) Charge-transfer-complex between the dye and the acceptor. (2) "Trap-action" of the acceptor.

of the holes in the fundamental band of the dye, and therefore the conductivity of the *p*-conductor. This is proved among other things by ESR measurements,[652,664,668] which reveal an increase in conductivity with increase in the number of electrons which have accumulated in the acceptor molecules on increasing the concentration of the doping agent. Even the reverse possibility of the doping of *n*-conducting organic compounds (acceptors) with electron donors[55,665] is in accordance with the hypothesis.

Quantum yield $G > 1$: quantum yields of $G > 1$ might be amenable to an analogous explanation to that given in the case of inorganic photoconductors, e.g. CdS.[215,332] This signifies that the electrons which have accumulated in the acceptors and which for the most part are immobile cause a negative space charge—corresponding to the diagram in Fig. 46—to be set up at the anode during exposure, and under the influence of this space charge, positive holes are injected from the anode. Arguments in favour of the last-named process are supported by the studies made by *Kallmann, Pope et al.*,[669-671] which give experimental proof of the injection of holes from electrodes into organic photoconductors of the *p*-type, e.g. into anthracene.

Because of this after-delivery effect on the part of the positive holes, a positive hole is also virtually present even after it has entered the cathode. Since this is formally equivalent to an extension of the lifetime τ of the charge

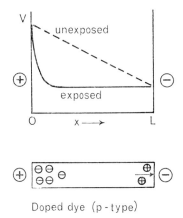

Fig. 46. Diagrammatic representation of the space charge relations in an unexposed and an exposed doped *p*-conducting dye.

carrier, the quantum yield (gain factor) G which is defined by equation 69 can become greater than $1 (\eta \to 1)$ when $\tau > T$.

$$G = \eta \cdot \frac{\tau}{T} \tag{69}$$

Equation 69 is obtained by inserting in equation 52 the time T taken by a charge carrier to travel from one electrode to the other

$$T = L/\mu_e \cdot E \tag{70}$$

It should also be noted that not only is the effect of doping of importance in connection with supersensitization, but that it also indicates which processes must be employed in order to be able to obtain organic photoconductors of possible future industrial use.

6.2.2.6 *The Connection between Photoconductance and Desensitization.* Any comparison of the photoconducting properties of sensitizing and desensitizing dyes is only possible when it is limited to dyes of similar structure. This is because distinct differences in constitution such as those which occur for example between non-ionic merocyanines and cationic triphenylmethane dyes lead one from the outset to expect not only differences in their conductivity behaviour but also in their photographic usefulness (e.g. on account of anomalous adsorption power, etc.) and these differences would hamper any discussion.

A few dyes of similar structure which appear to be suitable for making a comparison are compiled in Table 18.

TABLE 18

Comparison	Dye	Photographic behaviour	Abs. max	Conductivity
1. 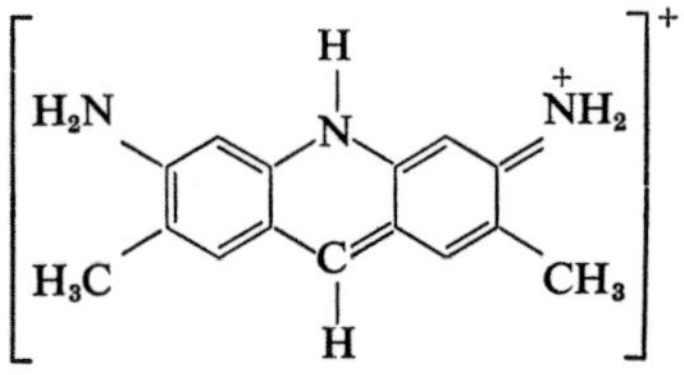 [69]	Acridine yellow	sensitizer	456 mμ	p-type
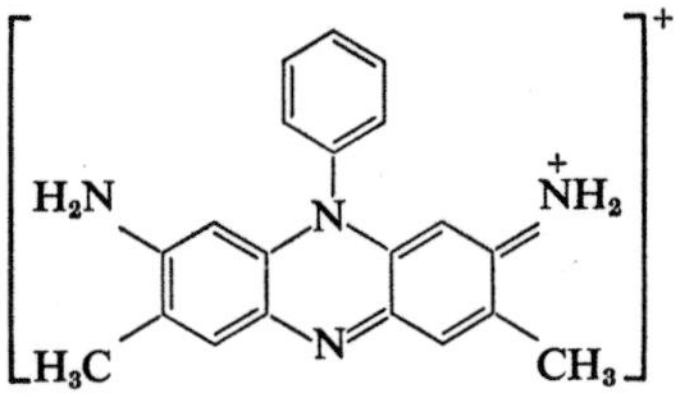 [70]	Safranine	desensitizer	539 mμ	n-type
2. 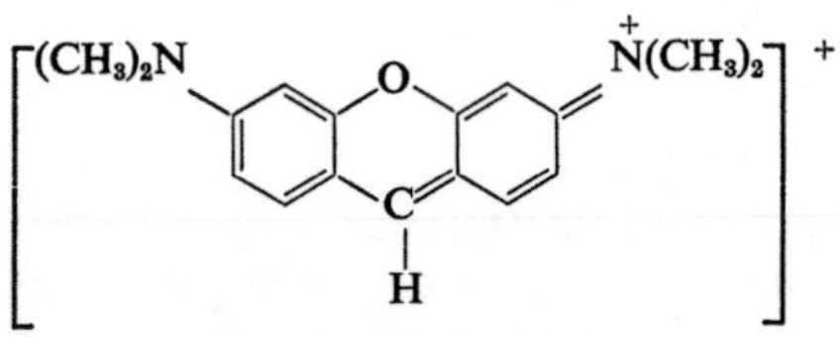 [71]	Pyronine G	sensitizer	550 mμ	p-type
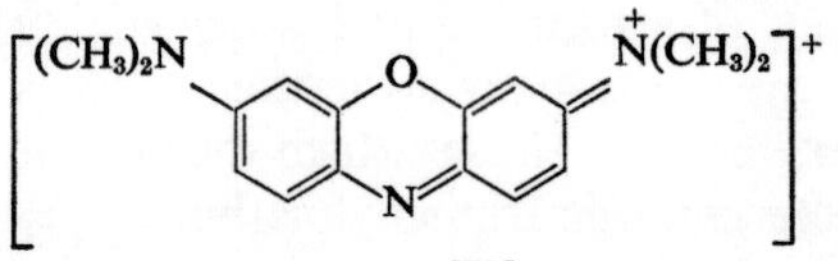 [72]	Capriblue	desensitizer	665 mμ	n-type

TABLE 18 (*contd.—*)

Comparison	Dye	Photographic behaviour	Abs. max	Conductivity
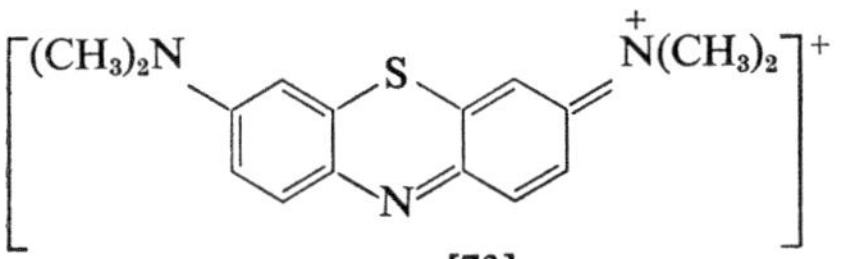 [(CH₃)₂N—⟨⟩—S—⟨⟩=N(CH₃)₂]⁺ [73]	Methylene blue	desensitizer	667 mμ	*n*-type
3. 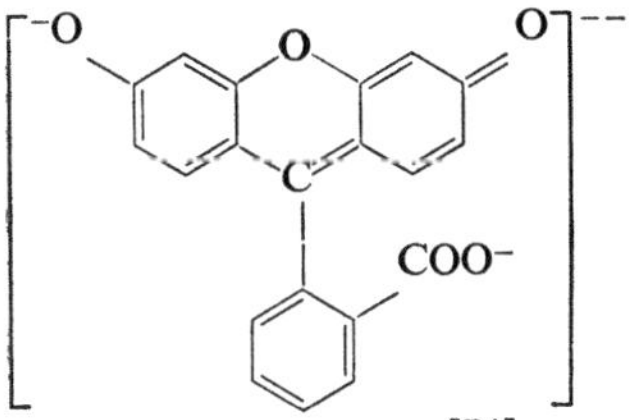 [⁻O—⟨⟩—O—⟨⟩=O]⁻⁻ with COO⁻ [74]	Fluorescein	sensitizer	490 mμ	*p*-type
4. 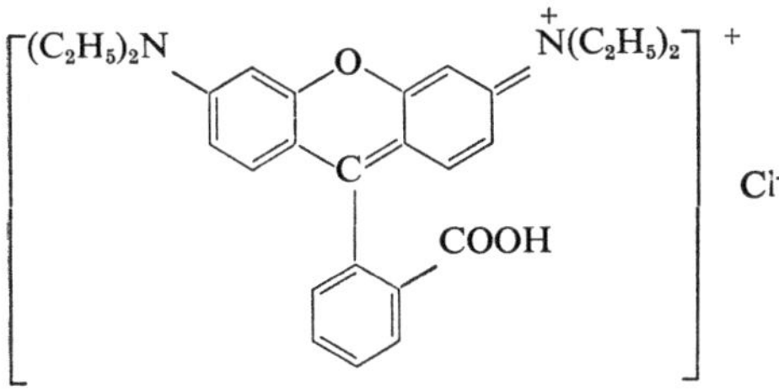 [(C₂H₅)₂N—⟨⟩—O—⟨⟩=N(C₂H₅)₂]⁺ Cl⁻ with COOH [75]	Rhodamine B	desensitizer (but also sensitizer)	550 mμ	*n*-type

From the examples 1 and 2 it can be seen that replacement by a nitrogen atom, of the bridging carbon atom which has been positively induced by the auxochrome groups, not only displaces the maximum absorption by about 100 mμ towards the long wavelengths but that it also produces a desensitizing action (see Chap. II).

The long wavelength displacement can be explained in accordance with the electron gas model in that the N atom, because of its greater electron affinity as compared with the methine group, exerts a sucking action on the electrons in the vibrating electron gas in the conjugation chain, thereby causing the excitation level to sink, since it is precisely in the excitation state that the density distribution of the electrons in the given dyes is concentrated at the N atom.[53] The ground state is not affected by the insertion of the hetero atom because the electron density is a minimum at the central atom.

It is a remarkable fact that not only do the last-named dyes show desensitizing properties, but they also behave like *n*-type organic photoconductors, whereas the sensitizing compounds of analogous constitution which have a C atom instead of the central N atom, belong to the *p*-conductors. Since the comparison is limited to dyes of similar constitution, the foregoing statement must not be taken as a general rule, for example, it cannot be said that sensitizers are generally *p*-conductors and that desensitizers are *n*-conductors. When grading *n*- and *p*-conducting dyes it is also necessary amongst other things to take into account the influence of cations or anions which contain vacant sites and which can act as donors or acceptors (e.g. in Rhodamine B and Fluorescein*, respectively, in Table 18). The importance of this observation lies more in the fact that it may give an indication of the mechanism of desensitization and of the creation of *n*- and *p*-conducting types of those organic photoconductors in which there is a falling off in the influence of cationic or anionic lattice defects.

The desensitizing properties of dyes produced by the introduction of an N atom indicate e.g. that a desensitizing action can be obtained in dyes of similar constitution by increasing their electron affinity. Dyes with strongly electron-attracting groups—e.g. nitro groups—therefore represent powerful desensitizers from the very outset.

A striking fact is that this behaviour is accompanied by a change in the photoconductivity, caused most probably by an increase in the electron affinity of the bridging atom and not by anomalies in constitution which are difficult to detect (differences in the superposition of the molecules, etc.).[672] How far this change is linked with the relation which exists between log σ_D and the

Y' and Y are respectively —OCH_3, —CH_3, CL, —Br, —NO_2

[76]

Hammett constants* of the *m*- and *p*-substituents—decrease in the dark conductivity with increase in the Hammett constant—and which was observed with differently substituted anils[673] remains uncertain. If *p*-conduction is assumed to take place in anils [76] the decrease in the dark conductivity brought

* By the Hammett constant is understood a constant which is derived from the dissociation equilibrium of substituted benzoic acids, and which specifies the relative electron density for various substituents which are situated at the sites of reaction in the *m*- or *p*-position to the substituent in a molecule.

about by more powerful electron-attracting substituents (Hammett constant: $CH_3 < H < OCH_3 < Cl < Br < NO_2$) could be attributed to an increasing retardation in the intermolecular transport of holes with the positivation of the N atom. On applying this hypothesis to dyes in which positivation of the bridging atom is not produced by changing the substituents but by replacing the C atom by an N atom, a decrease in the mobility of the positive holes would, for analogous reasons, likewise be a debatable point. Since at the same time in consequence of the sinking of the excitation states there is an increased probability of the overlapping of the π-electron clouds of the excitation states, in these compounds, on the contrary, an increase in the mobility of the π-electrons in the conduction band has to be reckoned with, i.e. substitution can produce an alteration in the ratio $\mu_\oplus : \mu_\ominus$.

The correctness of this assumption is possibly indicated by the mobilities of $\mu_\oplus$ and $\mu_\ominus$ as measured by *Kepler*[597] in the case of anthracene: perpendicular to the *ab*-plane $\mu_\oplus = 0.4$ (cm^2 . V^{-1} . sec^{-1}), $\mu_\ominus = 0.3$ (cm^2 . V^{-1} . sec^{-1}), whereas parallel to the *ab*-plane $\mu_\oplus = 1.3$ (cm^2 . V^{-1} . sec^{-1}) and $\mu_\ominus = 2.0$ (cm^2 . V^{-1} . sec^{-1}). The observation of *n*-conduction in a Pinakryptol green desensitizer containing an NO_2 group would also conform with this hypothesis.

In summarizing it may be said that in any case these results reveal that the conductivity behaviour of pure sensitizing or desensitizing dyes is connected with the photographic properties of the dyes. This is also indicated by the work carried out by *Wilk*,[674] that is, the author was able to prove that the empirical relation given by equation 71 between conductivity (expressed by I_{phot}/I_{dark}) and the polarizability which was measured from the displacement of the absorption bands ($\Delta\lambda$[kcal]) which occurred on changing the solvent, was valid for sensitizers but not for desensitizers.

$$I_{phot}/I_D = k \cdot (\Delta\lambda)^2 - 1 \qquad (k = 1.3) \qquad (71)$$

Moreover, this conclusion seems to be confirmed by the following fact: Desensitizing dyes (which often belong to *n*-type photoconductors) in many cases show enhanced photoconductivity with decrease in their oxygen content[485] and in agreement with this observation is the fact that during evacuation Phenosafranine shows not only a marked decrease in its desensitizing action but also a marked spectral sensitization effect.[647a] The differences in the influence of oxygen on dyes led several years ago to the classification of dyes into sensitizers and desensitizers and to the postulate that oxygen can to some extent prevent photoelectrons from being injected into silver halide. Cf. [53,p.189,485] The recent measurements made by *Lewis* and *James*[674a] appear to be in good agreement with this conclusion, but more work seems to be necessary before any further discussion can take place.

6.2.3 THE QUESTION OF THE ENERGIES INVOLVED IN SPECTRAL SENSITIZATION.

6.2.3.1 *Definitions.* The question of the absolute positions of the energy levels in organic dyes is at the present moment still open to discussion. The assumption which runs contrary to the electron transfer hypothesis of sensitization, namely that the ground level in a dye (for example in merocyanines) is situated 7·35 eV below the limit of ionization,[408] might however be considered to be positively out of date.

In considering the problem it is advisable to regard dyes as inorganic semiconductors and to use the same methods for determining the positions of the energy levels in both cases.

In analogy to inorganic solids, macroscopic space charges, double layers, etc., in an organic system will moreover also influence the energy appertaining to an electron, because of the periodic potential of the lattice. If, when considering this influence, the electrostatic potential or the macropotential ψ is introduced,[586] then the chemical binding energies of the electrons can be calculated exactly from this potential onwards, and the total energy of an electron is therefore given by (cf. the discussion in[53]):

$$E_{\text{total}} = E - e\psi \qquad (72)$$

E (e.g. E_L, E_D) < 0 and $\psi \lessgtr 0$.

Every change in ψ which is caused by any kind of charges (adsorption, etc.) produces analogous displacements *in all* the electron levels. When considering for instance the regeneration process during sensitization it would therefore be unreasonable to expect that any explanation which is based on differences in the displacement of the energy bands of a silver halide[675] could fall into line with the actual circumstances.

Since the quantities given in the literature for the energies of solid organic dyes are inconsistent—cf. for example [443,646]—a brief synopsis of these quantities will be given here (see also Fig. 47).

The Thermal Work Function W_{therm}: The thermal work function W_{therm} of electrons is defined as that energy which must be expended in order to effect the thermal release of an electron from a solid. W_{therm} therefore corresponds to the difference in energy between the electrostatic energy $-e\psi$ (which may be influenced by a jump in the potential D of the double layer) and the Fermi energy E_F.

The work function of electrons can be determined from the contact potential (*e* . voltage) of two substances *in vacuo* if the work function of one of these solids is known. Two methods are used for testing dyes. Firstly, *Thomson*'s vibration method in which the voltage set up at one oscillation condenser is compensated by a counter-voltage.[408] Secondly, the method in which the current voltage characteristic is recorded during the photoelectric—cf. Chap. V, section 3.8[585]—or thermal release of electrons.[646]

In this connection it should be pointed out that the data of the electron affinities of a dye were taken from values of the current voltage characteristics during the thermal release of electrons from a reference electrode, measured against layers of the dye.[646] This interpretation might only be possible however in the case of n-conductors, the Fermi level E_F of which lies close to the edge of the conduction band E_c.

The Fermi Energy E_F (Chemical and Electrochemical Potential, respectively): If disturbances due to adsorbed gases, etc., are excluded, W_{therm} corresponds to the Fermi energy E_F. A knowledge of this energy quantity—apart from data of the occupation of the level—is of great importance, above all for describing systems in contact. It must not however be overlooked that on

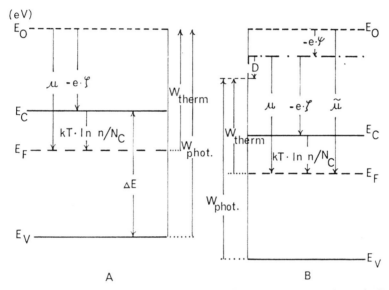

Fig. 47. Diagram of the energy bands (Notation, see text). A: Macropotential $\psi = 0$. B: Positive charging up of the surface ($\psi > 0$).

account of the difficulty of measuring W_{therm} in the case of dyes (due amongst other things to the influence of adsorbed gases[676] which cannot be removed here by heating) the values given in the literature for the Fermi energies of dyes show large discrepancies.[677] From section 5.3.8 it can however be seen that the Fermi level of exposed, n-conducting dyes is consistent with theory in that it is situated in the neighbourhood of the conduction band and is displaced towards the fundamental band in the case of p-conducting dyes.

It should also be noted that the Fermi energy corresponds to the chemical potential μ of the charge carriers[678]

$$(E_F \equiv)\mu = E_c + kT \ln n/N_c \quad [\text{eV}] \tag{73}$$

and to the electrochemical potential (with reference to E_0) respectively when an electrostatic constituent of the energy (the macropotential) $-e\psi$ is present

$$(E_F \equiv)\tilde{\mu} = -e\psi + E_c + kT \ln n/N_c \quad [\text{eV}] \qquad (74)$$

where n is the concentration of the charge carriers, N_c is the effective density of states in the conduction band, and E_c is the energy of the conductivity band (< 0).

The Photoelectric Activation Energy W_{phot}: By the photoelectric activation energy is understood that energy which must be expended for the photoemission of an electron, and which can be determined from the long wavelength limit of the external photoeffect. Whereas in a metal W_{phot} corresponds with W_{therm}, this is certainly not true of a dye, because here the photoelectric emission from the lattice defect levels[589,679] or from a fully occupied valency band[585] does not take place from a Fermi level, if the case of a dye photoconductor containing a very large number of lattice defects be excluded, when E_F would coincide with the levels of the lattice defects. When electrons are emitted from the valency band, W_{phot} is identical with the energy of ionization I and can therefore be used for determining the position of the valency band E_V.

Although a series of investigations has already been carried out on the energies of ionization of organic compounds,[646,573,677,679-690] the values which have hitherto been given in the literature often show large discrepancies. Further measurements are therefore necessary.

It should also be mentioned that *Nelson*,[691] in the course of measurements of the energies of ionization of dyes, observed that the positions of the energy levels of layers of the dyes coincided with those of the individual dye molecules.

The Electron Affinity $-e\zeta$: The electron affinity $-e\zeta$ is identical with the energy E_c of the lower edge of the conduction band of a dye. As already mentioned, *Nelson*'s method[646] (namely the measurement of thermal current–voltage characteristics) enables some information to be obtained about the electron affinity in special cases. The values of electron affinity derived from the difference between the photoelectric activation energy W_{phot} and the distance between the energy bands ΔE in accordance with equation 75, can be regarded as giving a better approximation:

$$-e\zeta = W_{\text{phot}} - \Delta E \quad [\text{eV}] \qquad (75)$$

Measurements of ζ via the polarographic cathodic half-stage potential[260,288,692,693] might be more suitable for control measurements within homologous series.

The Distance between the Energy Bands ΔE: The distance between the fundamental and the conduction bands $\Delta E = E_C - E_V$ which is of importance amongst other things for determining the electron affinity in solid dyes, can be obtained firstly from the absorption spectrum by measuring the limiting wavelength λ_g. Fig. 48 shows how this can be done using a layer of Crystal violet as an example.

Secondly, the long wavelength limit $\lambda_g{}^*$ of the photoconductivity also leads to ΔE if the absorption of lattice defects is excluded as being the cause of the photoconductance.

On account of many influences which are difficult to detect (variations in the energies of solvatation and in the rates of diffusion of the redox products, and also in the electron affinity of the electrodes and again in the diffusion potential at the boundary phase to the reference electrode when organic

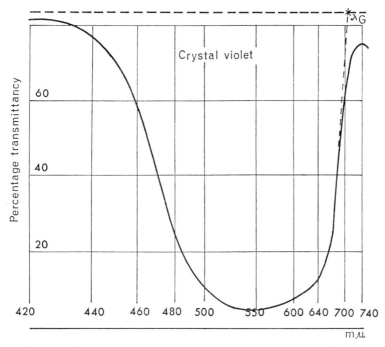

Fig. 48. Crystal violet. Transmittance in relation to wavelength. (Thickness of the layer 10^{-4} cm.)

solvents are used, etc.) the electrochemical half-stage potentials of dyes cannot be compared directly with the quantities of energy measured *in vacuo*, nevertheless in homologous series of dyes a connection exists between these potentials and the difference in energy ΔE. This is probably another consequence of the energy levels in the polymethine cyanine systems being influenced largely in an analogous way. According to *Stanienda*[693,693a,693b] a linear relation exists between the difference between the anodic and cathodic half-stage potentials of polymethine dyes $\Delta E_{\frac{1}{2}}$ which corresponds in principle with the difference in the energy of the highest occupied and the lowest free electron

173

levels, and the distance of the energy level ΔE which can be derived from the absorption limit. The following holds:

$$\Delta E_{\frac{1}{2}} = a \cdot \Delta E - n \tag{75a}$$

where $a = 1\cdot36$; $b = 1\cdot25$ for thiapolymethine cyanines, and $a = 1\cdot35$; $b = 1\cdot07$ for 2,2'-quinapolymethine cyanines, respectively[41] (in acetonitrile).

TABLE 19

QUANTITIES OF ENERGY OF SOME ORGANIC DYES

Dye	Type	E_F (W_{therm}) [eV]	E_V (W_{phot}) [eV]	E_C ($-e\zeta$) [eV]	ΔE [eV]	Literature
Rhodamine B	n	4·45	5·2	3·2	2·0	402, 408
		4·6	5·1	—	—	688
		3·5*	—	—	2·0	
Erythrosine	p	4·64	5·5	3·3	2·2	408
		—	5·5	3·4	2·0	688
		4·5*	—	—	2·0	
Malachite green	n	4·86	—	—	2·0	408
		4·6	5·2	—	—	688
		—	5·2	3·2	1·6	687
		3·8*	—	—	1·6	
Pinacyanol	n	4·55	4·9	3·1	1·8	408
		—	7·28	5·1	—	408
		5·3	5·6	3·3	2·3	688
		3·5*	—	—	1·6	
Crystal violet	n	3·32	5·1	3·0	1·7	513, 650, 690
		4·5	5·0	—	—	513
		3·8*	—	—	1·7	
Merocyanine	p	—	5·8	3·5	2·3	408
		—	7·35	4·8	—	408
		5·3	5·6	3·3	2·3	688
Merocyanine FX 79 [60]	p	4·5*	—	—	1·9	
Phthalocyanine	p	—	6·0	4·3	1·7	408
		4·5	—	—	1·7	
Phenosafranine	n	5·0	5·4	3·3	2·1	688
		3·9*	—	—	—	
Methylene blue	n	5·2	5·4	3·6	1·8	688

* Values for exposed dyes.

Further experiments are necessary in order to ascertain the extent to which this relation is generally valid; see also cf. [693c] and [693d].

6.2.3.2 *Previous Results*. Table 19 gives a survey of a few of the energy quantities which have previously been obtained for dyes.

The table shows the following:

1. In conformity with *Scheibe*'s hydrogen rule,[281,640,641,694,695] the lower edge of the conductivity band of layers of dye (given by the electron affinity) is about 3 to 3·5 eV distant from the ionization limit. This proves that *Scheibe's* rule

TABLE 19A

QUANTITIES OF ENERGY OF SOME SEMICONDUCTORS

Semiconductor	Type	E_F (W_{therm}) [eV]	E_V (W_{phot}) [eV]	E_C ($-e\zeta$) [eV]	ΔE [eV]	Literature
AgBr	(p)	—	6·0	3·5	2·5	402, 408, 513, 685 693e
TlI	p	5·4	5·6	3·0	2·5	251, 408
CdS	n	—	—	3·5	—	513
ZnO	n	4·84	7·3	4·3	3·0	402, 408
		—	6·3	—	—	686

holds not only for individual dye molecules, but also for aggregates of molecules in layers.

2. A comparison with the energy levels in AgBr and other semiconductors

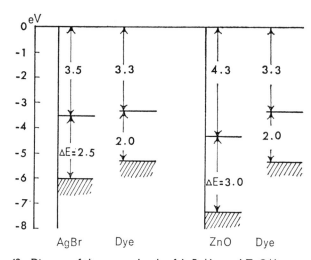

Fig. 49. Diagram of the energy bands of AgBr/dye and ZnO/dye systems.

(cf. Fig. 49) reveals that the first excitation level or conductivity band of dyes lies above or in the neighbourhood of the conductivity band of AgBr, TlI, CdS and ZnO. Consequently the position of the excitation level of a dye, which *Coulson et al.* assumed to lie about 4–6 eV below the ionization boundary, i.e. below the AgBr-conduction band, and which presented an argument against the transfer of electrons from the dye to the semiconductor and led to the relinquishment of the electron transfer theory, does not correspond with the conditions obtaining in dyes. This means that a transfer of electrons from a dye to AgBr and other semiconductors (CdS, etc.) is energetically possible.

3. The values of the Fermi levels of individual dyes as given by the thermal work function are greatly scattered, so that further measurements are necessary in order to establish the value of E_F accurately.

References

[610] SHEPPARD, S. E., *Chem. Rev.*, **4**, 340 (1927).
[611] BERTHOUD, A., *Z. phys. Chem.*, **120**, 174 (1926).
[612] SCHEIBE, G., *Naturwissenschaften*, **25**, 795 (1937).
[613] SCHEIBE, G., *Z. Elektrochem.*, **56**, 723 (1952).
[614] FRANK, J. and TELLER, E., *J. chem. Phys.*, **6**, 861 (1938).
[615] KREBS, A., *Ergebn. exakt. Naturw.*, **27**, 361 (1953).
[616] SCHMILLEN, A., *Z. Phys.*, **135**, 294 (1953); **150**, 123 (1958).
[617] SCHMILLEN, A. and KOHLMANNSPERGER, J., *Z. Naturf.*, **18a**, 627 (1963).
[618] FÖRSTER, TH., *Z. Elektrochem.*, **64**, 157 (1960).
[619] FÖRSTER, TH., *Ann. Phys.*, **2**, 55 (1948).
[620] WEIGERT, F., *Trans. Faraday Soc.*, **36**, 1033 (1940).
[621] WINTERSTEIN, A. and SCHÖN, K., *Naturwissenschaften*, **22**, 237 (1934).
[622] BOWEN, E. J., *J. chem. Phys.*, **13**, 306 (1945).
[623] GAUGULEY, S., *J. chem. Phys.*, **13**, 128 (1945).
[624] ZWICK, M. M. and KUHN, H., *Z. Naturf.*, **17a**, 411 (1962).
[625] DREXHAGE, K. H., ZWICK, M. M. and KUHN, H., *Ber. Bunsenges. phys. Chem.*, **61**, 62 (1963).
[626] KUHN, H., *Optische Anregung organischer Systeme* (*Optical Excitation of Organic Systems*), edited by W. FOERST, p. 639, Verlag Chemie (Weinheim/Bergstr.) 1966.
[627] PERRIN, J., *Ann. chim. Phys.*, **17**, 283 (1932).
[628] WAWILOW, S. I. and PHEOFILOV, P. P., *Dokl. Akad. Nauk SSSR*, **34**, 220 (1942).

[629] FÖRSTER, TH., *Fluoreszenz organischer Verbindungen* (*Fluorescence of Organic Compounds*), Vandenhoeck and Ruprecht (Göttingen) 1951.
[629a] BENNETT, R. G. and KELLOGG, R. E., *Progress in Reaction Kinetics*, Vol. 4., p. 217, Pergamon Press (Oxford, New York) 1967.
[630] COULSON, C. A., and N. F. MOTT, *Photogr. J.*, Ser. B, **88**, 119 (1948).
[630a] AKIMOV, I. A., BENTSA, V. M., VILESOV, F. I. and TERENIN, A. N., *Phys. stat. sol.*, **20**, 771 (1967).
[631] MITCHELL, J. W., *J. photogr. Sci.*, **5**, 49 (1957); **6**, 57 (1958).
[632] URBACH, F., *Phys. Rev.*, **92**, 1324 (1953).
[633] BILTZ, M., *J. opt. Soc. Amer.*, **39**, 994 (1949).
[634] EGGERT, J., *Ann. Phys.*, **7**, 4, 140 (1959).
[635] STASIW, I. and TELTOW, J., *Z. wiss Photogr.*, **40**, 157 (1941).
[636] STASIW, O., *Elektronen- und Ionenprozesse in Ionenkristallen* (*Electronic and Ionic Processes in Ionic Crystals*), Springer Verlag (Berlin) 1959.
[637] WEST, W., *J. Chim. phys.*, **55**, 672 (1958).
[638] BILTZ, M. and WEBB, J. H., *J. opt. Soc. Amer.*, **38**, 561 (1948).
[638a] ING, S. W. and CHIANG, Y. S., *J. chem. phys.*, **46**, 478 (1967).
[638b] BÜCHER, H., KUHN, H., MANN, B., MÖBIUS, D., VON SZENTPÁLY, L. and TILLMANN, P., *Photogr. Sci. Eng.*, **4**, 233 (1967).
[639] NELSON, R. D., DREXHAGE, K., SCHÄFER, F. and KUHN, H., *J. phys. Chem. Solids*, **27**, 1189 (1966).
[639a] HARTMANN, W., *Z. techn. phys.*, 17, 436 (1939).
[639b] VAN GEEL, W. CH. and DE BOER, J. H., *Physica*, **2**, 892 (1935).
[639c] HAMMOND, G. S., *Kagaku to Kôgyô* (*Tokyo*), **18**, 1464 (1965).

[639d] HERKSTROCKTER, W. G., LAMOLA, A. A. and HAMMOND, G. S., *J. Amer. chem. Soc.*, **86**, 4537 (1964).
[639e] STEINMETZ, R., *Fortschr. chem. Forsch.*, 7/3, 445 (1967).
[639f] LAMOLA, A. A. and HAMMOND, G. S., *J. chem. phys.*, **43**, 2129 (1965).
[640] SCHEIBE, G., BRÜCK, D. and DÖRR, F., *Ber. dtsch. chem. Ges.*, **85**, 867 (1952).
[641] SCHEIBE, G. and BRÜCK, D., *Z. Elektrochem.*, **54**, 403 (1950).
[642] SCHEIBE, G., *S. B. Bayer Akad. Wiss., München, math.-nat. Kl.*, 11th Jan. 1952; *Angew. Chem.*, **63**, 174 and 439 (1951).
[643] NODDACK, W., ECKERT, G. and MEIER, H., *Z. Elektrochem.*, **56**, 735 (1952).
[644] MEIER, H., *Z. wiss. Photogr.*, **50**, 20 (1955).
[645] MATEJEC, R., *Z. Elektrochem.*, **61**, 281 (1957).
[646] NELSON, R. C., *J. opt. Soc. Amer.*, **46**, 1016 (1956).
[647] NELSON, R. C., *J. opt. Soc. Amer.*, **48**, 948 (1958).
[648] NELSON, R. C., *J. chem. Phys.*, **27**, 864 (1957).
[649] KLEIN, E. and MATEJEC, R., *Optische Anregung organischer Systeme (Optical Excitation of Organic Systems)*, edited by W. FOERST, p. 698, Verlag Chemie (Weinheim/Bergstr.) 1966.
[650] BLOCHINZEW, D. J., *Grundlagen der Quantenmechanik (Principles of Quantum Mechanics)*, p. 349, Deutscher Verlag der Wissenschaften (Berlin) 1953.
[651] NELSON, R. C., *J. phys. Chem.*, **69**, 714 (1965); *Vortragsmanuskript (Lecture Notes)*, Chicago, 22nd June, 1964, Intern. Conf. on Photosensitization in Solids.
[652] KEARNS, D. R. and CALVIN, M., *J. Amer. chem. Soc.*, **83**, 2110 (1961).
[653] MEIER, H. and ALBRECHT, W., *Z. phys. Chem.*, (N.S.), **39**, 249 (1963).
[654] KEARNS, D. R., TOLLIN, G. and CALVIN, M., *J. chem. Phys.*, **32**, 1020 (1960).
[655] AKAMATU, H., INOKUCHI, H. and MATSUNAGA, Y., *Nature, Lond.*, **173**, 168 (1954); *Bull. chem. Soc., Japan*, **29**, 213 (1956).
[656] AKAMATU, H. and INOKUCHI, H., *Symposium on Electrical Conductivity in Organic Solids*, p. 277, Interscience Pub. (New York) 1961.
[657] BRIEGLEB, G., *Elektronen-Donator-Acceptor-Komplexe, (Electron-Donor-Acceptor-Complexes)*, Springer (Berlin-Göttingen-Heidelberg) 1961.
[658] DUNNING, N. J., *Dielectrics*, **1**, (4), 201 (1964).
[659] NISHIZAKI, S. and KUSAKAWA, H., *Bull. chem. Soc., Japan*, **36**, 1681 (1963).
[660] AKAMATU, H. and KURODA, H., *J. chem. Phys.*, **39**, 3364 (1964).
[661] MATSUNAGA, Y., *Helv. phys. acta*, **36**, 800 (1963).
[662] KOKADO, H., HASEGAWA, K. and SCHNEIDER, W. G., *Canad. J. Chem.*, **42**, (5), 1084 (1964).
[663] LAPINSKI, J. H. and HUGGINS, C. M., *J. phys. Chem.*, **66**, 2221 (1962).
[664] OTTENBERG, A., HOFFMAN, C. J. and OSIECKI, J., *J. chem. Phys.*, **38**, 1898 (1963).
[665] REUCROFT, P. J., RUDYI, O. N., SALOMON, R. E. and LABES, M. M., *J. phys. Chem.*, **68**, 779 (1965).
[666] MULLIKEN, R. S., *J. Amer. chem. Soc.*, **72**, 600 (1950); **74**, 811 (1952).
[667] MEIER, H. and ALBRECHT, W., *Angew. Chem.*, **77**, 177 (1965).
[668] KOHIN, R. P., MÜLLER, K. A. and HOEGL, H., *Helv. phys. acta*, **35**, 255 (1962).
[669] KALLMANN, H. and POPE, M., *Nature, Lond.*, **186**, 31 (1960); *J. chem. Phys.*, **32**, 300 (1960); *Symposium on Electrical Conductivity in Organic Solids*, edited by H. KALLMANN and M. SILVER, p. 1, Interscience Pub. (New York) 1961.
[670] SILVER, M. and MOORE, W., *J. chem. Phys.*, **33**, 1671 (1960).
[671] SILVER, M., *Organic Semiconductors*, edited by J. J. BROPHY and J. W. BUTTREY, p. 27, Macmillan (New York) 1962.
[672] CARLTON, D. M., MCCARTHY, D. K. and GENZ, R. H., *J. phys. Chem.*, **68**, 2661 (1964).
[673] GOODEN, E. W., *Nature, Lond.*, **203**, 515 (1964).
[674] WILK, M., *Z. Elektrochem.*, **64**, 294 (1960).
[674a] LEWIS, W. C. and JAMES, T. H., presented at Int. Congress of Photographic Science, Tokyo, (1967).
[675] KLEIN, E., MATEJEC, R. and MOLL, F., *Scientific Photography*, Proc. Int. Colloq., Liége 1959, edited by H. SAUVENIER, p. 569, Pergamon Press (Oxford) 1962.
[676] REDFIELD, D., *Appl. Physics Letters*, **1**, 9 (1962).
[677] FOX, D., LABES, M. and WEISSBERGER, A., *Physics and Chemistry of the Organic Solid State*, Vol. I, Interscience Pub. (New York) 1963.
[678] GÖHR, H. F., *The Electrochemistry of Semiconductors*, edited by P. J. HOLMES, p. 1, Academic Press (London) 1962.
[679] NELSON, R. C., *J. molecular Spectroscopy*, **7**, 439 (1961).
[680] WEST, D. C., *Canad. J. Phys.*, **31**, 691 (1953).

[681] LYONS, L. E. and MACKIE, J. D., *Proc. chem. Soc., Lond.*, 71 (1962).
[682] KEARNS, D. and CALVIN, M., *J. chem. Phys.*, **29**, 950 (1959).
[683] BECKER, R. S. and WENTWORTH, W. E., *J. Amer. chem. Soc.*, **85**, 2210 (1963).
[684] WATANABE, H., *J. chem. Phys.*, **22**, 1565 (1954); **26**, 542 (1957).
[685] TAFT, E. A., PHILIP, H. P. and APKER, L., *Phys. Rev.*, **110**, 876 (1958).
[686] VILESOV, F. I. and TERENIN, A., *Dokl. Akad. Nauk SSSR*, **125**, 55 (1959); **133**, 1060 (1960); **134**, 71 (1960); *Naturwissenschaften*, **46**, 167 (1959).
[687] VILESOV, F. I., *Dokl. Akad. Nauk SSSR*, **132**, 632 (1960).
[688] KURBATOV, B. L. and VILESOV, F. I., *Dokl. Akad. Nauk SSSR*, **141**, 1343 (1961); *Soviet phys. Ber.*, **6**, 1091 (1962).
[689] FLEISCHMANN, R., *Ann. Phys.*, **5**, 73 (1930).
[690] NELSON, R. C., *J. opt. Soc. Amer.*, **51**, 1186 (1961).
[691] NELSON, R. C., *J. opt. Soc. Amer.*, **55**, 897 (1965).
[692] MATSEN, F. A., *Proceedings of the 1957 Conference on Carbon*, p. 27, Pergamon Press (New York) 1957.
[693] STANIENDA, A., *Naturwissenschaften*, **52**, 105 (1965).
[693a] STANIENDA, A., *Naturwissenschaften*, **47**, 353 (1960); **47**, 512 (1960).
[693b] STANIENDA, A., *Z. physik. Chem.*, **32**, 238 (1962); **33**, 170 (1962).
[693c] TANI, T. and KIKUCHI, S., *Photogr. Sci. Eng.*, **11**, 129 (1967).
[693e] PETERSON, C. W., *Phys. Rev.*, **148**, 335 (1966).
[693d] TANI, T. and KIKUCHI, S., *Kogyo Kagaku Zasshi*, **69**, 2053, 2237 (1966).
[694] SCHEIBE, G., *Chimia.*, **15**, 10 (1961).
[695] SCHEIBE, G., KERN, J. and DÖRR, F., *Z. Elektrochem.*, **63**, 117 (1959).

VII. THE BOUNDARY LAYER OR PN MECHANISM OF SPECTRAL SENSITIZATION

Even if the photoconductivity of organic dyes and the positions of the energy levels in sensitizer/semiconductor systems may be regarded as a proof that various part-reactions are involved in the electronic mechanism of sensitization originally discussed (the formation and migration of electronic charge carriers in the sensitizer, the physical possibility of the transfer of electrons), nevertheless a number of questions still remain open: for example, the fact that p-conductors can be sensitized, which is equivalent to a transfer of positive holes from the dye to the semiconductor, or why the dye is able to react several times (regeneration).

In order to elucidate these questions, an attempt was made to render the electronic processes which take place between a sensitizing dye and an inorganic semiconductor on exposure amenable to direct measurement by the use of models of sensitized systems consisting of combinations of inorganic and organic photoconductors, and thus to arrive at a theoretical explanation. The analogy between the electrical behaviour of organic and inorganic photoconductors (e.g. in regard to n- and p-conduction) suggested that combinations of n- or p-conducting dye layers with p- or n-conducting inorganic (or organic) photoconductors must give rise to photovoltaic effects on exposure as in the case of inorganic systems of the pn- or the nn'-type.

7.1 Investigations using Model Arrangements of Sensitized Systems

7.1.1 THE MEASURING ARRANGEMENT. The electrical behaviour of combinations of inorganic and organic photoconductors was tested using the arrangements shown diagrammatically in Fig. 50.[215,696]

The *first arrangement* consists of a semiconductor and a layer of dye placed in series next to one another between two electrodes set up on glass or quartz at a distance of about 0·3–1 mm from each other. The photovoltages and the photocurrents formed on exposing this arrangement were measured with an oscillation condenser electrometer and a galvanometer respectively of suitable sensitivity.

The high resistance of the dye components make it necessary to keep the distance between the semiconductor and the electrode which is bridged over by the dye, as small as possible—less than 0·2 mm. It should be noted that great difficulties are experienced in putting dyes from a solution—e.g. Malachite green, Pinacyanol, etc.—into the narrow gap between the semiconductor and the electrode: amongst other things it is impossible to prevent some of the inorganic material from being covered up. For this reason the experiments had to be confined in the first instance mainly to dyes which could be sublimated (phthalocyanines, rhodamines and merocyanines).

Undoubtedly this arrangement can only be regarded as a very much simplified model of a photoconductor which has been sensitized by the addition of a dye, but it does, however, enable semiconductor/dye combinations to be tested in a manner which can easily be supervised and with practically the entire elimination of unwanted boundary layers.

The *second arrangement* is a more faithful imitation of a sensitized semiconductor system than is the previous model, since the semiconductor layer and the layer

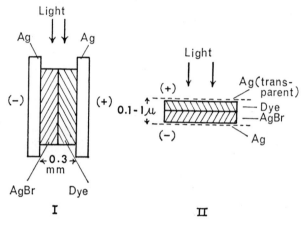

Fig. 50. Diagram of the measuring arrangements for determining the photovoltaic effect in inorganic/organic systems.

of dye which is about 0·1–1 μ in thickness, are arranged on top of each other as in a sensitized semiconductor. The charges which are formed in the light are led off by means of two semi-transparent electrodes (Ag, Pb) deposited by evaporation, and the short-circuit currents and the photovoltages are measured as described above with a sensitive galvanometer and an electrometer, respectively, and recorded by means of a recording instrument.

The only difference between this arrangement and spectrally sensitized systems—apart from the additional measuring electrodes—is that the layer of dye which is adsorbed to semiconductors in practice is usually thinner than that in the model, although sensitizing effects with dye-layers of the order of 1 μ are quite possible (cf. Fig. 18). Nevertheless it must not be assumed that the change to thinner layers which, however, are still multimolecular, produces a fundamental alteration in the processes which are observed to take place between the dye and the semiconductor in the model, for neither in photographic behaviour nor in conductivity was it possible to observe any fundamental changes on decreasing the thickness of the layer.

It is not until monomolecular layers of dye or groups of dye aggregates are used that it becomes impossible to exclude fundamental deviations from the outset, since the change from monomolecular to multimolecular layers alters the photographic behaviour of some dyes (particularly the cyanines) by replacing their sensitizing properties by desensitizing properties.

The photoconductors, layers of CdS and CdSe respectively ($0 \cdot 1$–1 μ), which were incorporated in the said arrangements could be produced by the evaporation of highly purified CdS and CdSe powders respectively without any special additions.[215] The silver bromide layers, on the other hand, were obtained in the manner described by *Eggert* and *Amsler*[343,344,378] by the electrolytic surface bromination of a silver layer immersed in a KBr solution (current density 30 mA/cm^2, 120 sec). The AgI layers were formed in principle by allowing iodine vapour to act on a coating of silver.[697,698] In addition to the inorganic/organic arrangements, tests were also carried out on the behaviour of combinations of pure organic dyes, since spectral sensitizing effects can likewise be observed in organic photoconductor systems. (See Chap. IV, section 3.)

7.1.2 THE RESULTS OF THE MEASUREMENTS. The following results were obtained with the said arrangements.

7.1.2.1 *Order of Magnitude of the Photovoltaic Effect.* On exposing various inorganic/organic systems (CdS/dye, AgI/dye, etc.) as well as combinations of *n*- and *p*-conducting dyes (Malachite green (*n*)/merocyanine (*p*), etc.)[699] (with an irradiation efficiency of 10^3 μ W/cm^2) photovoltages (U_∞) of the order of magnitude of 100–300 mV and photocurrents (I_0) of not exceeding 10^{-8} A are set up. Table 20 gives a compilation of a few of the systems studied, together with the maximum attainable photovoltages.

7.1.2.2 *Factors Governing the Directions of the Photocurrent and the Photovoltage.* The directions of the photocurrent and the photovoltage depend on the type of conduction in the combined photoconductors.

For instance, in CdS/merocyanine systems the dye (*p*-type), independently of the direction of the exposure, becomes positively charged and the CdS (*n*-type) negatively charged. When the combination contains an *n*-conducting dye the same polarity is observed, but the charge observed is smaller.

In silver halide/dye systems on the other hand the dye (if it belongs to the *n*-type) becomes negatively charged and the silver halide positively charged.

This direction of the charge which can be observed by measuring the photocurrent and the photovoltage respectively, indicates that an increase takes place in the concentration of the electrons in the CdS in CdS/dye systems, which is also a characteristic feature of the spectral sensitization of the photoconductivity of CdS. Measurements of silver halide/dye systems on the other hand show that electrons are transferred from the silver halide to the dye, i.e. that additional positive holes are formed in the silver halide as in the

TABLE 20

COMPILATION OF SOME OF THE DYE/SEMICONDUCTOR SYSTEMS STUDIED

Semiconductor (I)	Type	Dye	Type	Charge (I)	Max. U_∞ (10^3 $\mu W/cm^2$)	Note
CdS	n	Merocyanine A 10	p	—	250 mV	Formula, see Table 8 [59]
CdS	n	Merocyanine FX 79	p	—	200 mV	Formula, see Table 8 [60]
CdS	n	Cu-Phthalocyanine	p	—	220 mV	
CdS	n	Malachite green	n	—	80 mV	
					30 mV	according to Nelson[700]
CdSe	n	Merocyanine A 10	p	—	295 mV	[59]
AgBr	p	Rhodamine B	n	+	50 mV	[75]
AgBr	p	Merocyanine A 10	p	—	200 mV	[59]
AgI	p	Rhodamine B	n	+	350 mV	[75]
AgI	p	Malachite green	n	+	150 mV	
Malachite green	n	Merocyanine A 10	p	—	100 mV	[59]

case of the spectral sensitization of the photoconductivity of a silver halide (see Chap. IV). When, on the other hand, silver halides are brought into contact with *p*-conducting sensitizing dyes, the transfer of electrons takes place in the reverse direction, i.e. from the dye to the silver halide, as is perceived in photographic sensitization.

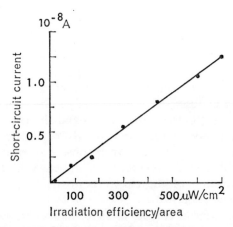

Fig. 51. CdS/Merocyanine. The short-circuit current as a function of the irradiation intensity ($\lambda = 5580$ Å).

7.1.2.3 Relation to Intensity. The short-circuit current I_0 and the photovoltage U_∞ increase with increase in the intensity of the irradiation. The photovoltage U_∞ shows a logarithmic and the short-circuit current I_0 a linear relation to the intensity of the irradiation I_λ (quanta irradiated per cm². sec).

$$U_\infty = A \cdot \log I_\lambda \tag{76}$$

$$I_0 = C \cdot I_\lambda \tag{77}$$

(A and C are constants and I_λ is the intensity of the irradiation)

as can be seen from the examples given in Figs. 51 and 52.

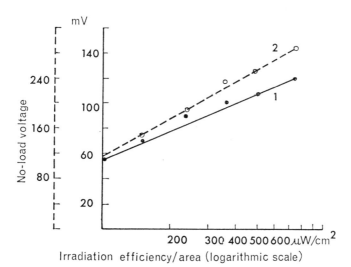

Fig. 52. The photovoltage as a function of the irradiation intensity. (1) CdS/Merocyanine A10 ($\lambda = 5580$ Å). (2) AgI/Rhodamine B ($\lambda = 6000$ Å).

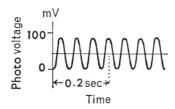

Fig. 53. CdS/Merocyanine A10. The photovoltage as a function of an intermittent exposure (irradiation efficiency 10^3 μW/cm²). Redrawn oscillogram.

7.1.2.4 Inertia. The photovoltage and the short-circuit current are formed rapidly in the various cells on exposure. The example shown in Fig. 53 is that of the oscillogram of a CdS/merocyanine photocell recorded with a sinusoidal intermittent exposure and redrawn. (Arrangement II in Fig. 50.) It can be seen that when the layers are in close contact, namely about 1 μ apart from each other, a photovoltage develops within the short space of time of $2 \cdot 10^{-2}$ sec. A correspondingly rapid development is also observed in pure *pn* dye systems, as is shown by the redrawn, recorded curve of the short-circuit current set up in the Malachite green (*n*)/Merocyanine A10 (*p*) cell in Fig. 54.

On the other hand, with distances of the order of magnitude of 0·01 cm between the electrodes, as in the first arrangement (Fig. 50), the rise and fall in the photocurrents and voltages is slower, and this is attributable to the inertia of the dye components and is apparently caused by traps.

7.1.2.5 *Spectral Sensitivity.* From the spectral distribution of the photosensitivity which is shown in Fig. 55, which gives an AgI/Merocyanine A10

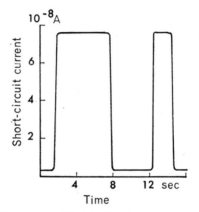

Fig. 54. Malachite green (*n*)/Merocyanine A10 (*p*). The short-circuit current as a function of the time of exposure. Irradiation efficiency $370 \cdot 10^{13}$ quanta per $cm^2 \cdot sec$ ($\lambda = 6220$ Å). Recorded curve redrawn.

cell as example, it is seen that the sensitivity of the semiconductor/dye systems lies farther towards the long wavelengths than that of the semiconductor (CdS, AgI, etc.) and this displacement can be attributed to absorption of light by the dye. This proves that the photovoltaic effect in semiconductor/dye arrangements, like the spectral sensitization of photoconductivity and of the photographic process, is produced by exposure to radiation of long wavelength which is absorbed by the dye. It is therefore natural to suppose that the various effects are combined.

7.1.2.6 *Current-voltage Characteristics.* The current-voltage characteristics of CdS/dye cells and of pure organic dye (*p*)/dye (*n*)-arrangements are in some cases unsymmetrical, as can be seen from the examples given in Fig. 56. When the polarity of the *n*-conducting components is positive the experimental values obtained for the photocurrent are greater than those which are obtained with

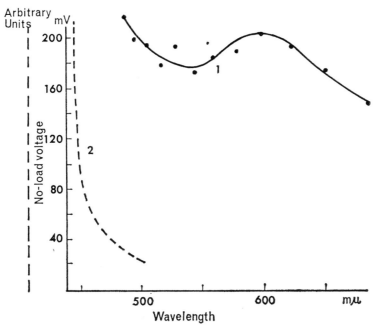

Fig. 55. (1) AgI/Merocyanine A10. The spectral distribution of the photovoltage for an irradiation efficiency/area of $10^3 \mu W/cm^2$. (Arrangement II of Fig. 50.) (2) AgI. The spectral distribution of the photosensitivity. After Terenin, et al.[402]

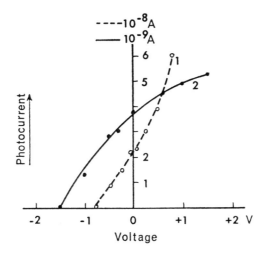

Fig. 56. The photocurrent as a function of the voltage. (1) Malachite green (n)/Merocyanine A10 (p). (Arrangement II of Fig. 50). (2) CdS/Merocyanine A10. (Arrangement II of Fig. 50.) Irradiation efficiency $10^3 \mu W/cm^2$ (4000–8000 Å).

the opposite polarity. Assuming that a boundary layer acts as a *pn*-junction between the *p*- and *n*-conducting compounds in contact, this observation would correspond to the action of a voltage laid on in the direction of the barrier and of the transmission respectively, at the *pn*-junction.[701]

In this connection it is interesting to note that layers of *p*- and *n*-conducting dyes—and indeed of *p*-conducting α-Cu-phthalocyanine and *n*-conducting β-chloro-indium-chlorophthalocyanine[580]—show a barrier action in the dark,[702] as when a *pn*-junction is present between the areas of a variable type of conduction. A barrier action is also found to exist between *p*- and *n*-conducting polyacrylnitrile[703] and is argued for the *pn*-junction in perylene complexes,[704] in organometallic polymeric semiconductor materials[705] and in doped polyethylene.[706] The absence of a corresponding rectifier action in similar systems must not, however, be taken as a contradiction, since a good barrier action cannot be achieved simply by pressing together pieces of the *p*-type and *n*-type even in the case of germanium.[707] As in this example, all kinds of unwanted influences have also to be reckoned with in combinations of inorganic and organic photoconductors, layers of which have been evaporated on top of one another. Such influences can give rise to a rapid recombination of the electrons and holes, and therefore to a decrease in the rectifier action.

It should be noted that rectifiers were constructed from organic semiconductors (phthalocyanines) and metal contacts in which the ratio of the transmission current/barrier current was up to 3000:1(10^4 Hz).[702,708–710]

7.1.3 INTERPRETATION OF THE RESULTS OF THE MEASUREMENTS.

7.1.3.1 *General Considerations.*

Neither the assumption of a diffusion photo-EMF which is due to falls in the concentration of the charge carriers brought about by variations in the absorption of light, nor an interpretation on purely photochemical lines is able to explain the photovoltaic effects observed. This is because the assumption of a diffusion-photo-EMF is contradictory to the experimental relation which exists between the direction of the photocurrents and the voltages, and a photochemical interpretation is ruled out since even extremely light-fast dyes (e.g. phthalocyanines) exhibit a photovoltaic effect when they are combined with semiconductors.

It should also be noted that when the semiconductors and dyes studied were built singly into the measuring arrangements (I or II in Fig. 50) they did not manifest any of the photovoltaic effects which occurred when they were in mutual contact.

For instance, the photocurrent which is characteristic of silver/silver halide systems—there is no photovoltage in these systems—and which was studied by *Eggert* and *Amsler*[343,344,378] in connection with the mechanism of the photographic primary process, superimposes itself only as a short, reversed part of the current on the short-circuit current of a silver/silver halide/dye (*n*)/silver arrangement at the commencement of the exposure. The low photovoltages which occur at dye/metal contacts cannot be brought into relation with the photoeffect in the dye/semiconductor systems, since, as Tables 20 and 21 show, the orders of magnitude of the photovoltages are not comparable.

This signifies that it is not the metal/semiconductor or the metal/dye contacts, but the boundary layer between the organic photoconductor and the dye (and between *p*- and *n*-conducting dyes, respectively) which is responsible for the photovoltaic effect in the model systems studied here.

7.1.3.2 *The pn theory of the Photovoltaic Effect in Model Systems.* The results obtained with semiconductor/dye systems (the relation between the directions of the short-circuit current and the photovoltage, the asymmetry

TABLE 21

THE PHOTO-EMF OF SYSTEMS: METAL/ORGANIC PHOTOCONDUCTORS

Compound	Material of the Electrodes	Photovoltage	Literature
Merocyanine A 10	Ag	1–3 mV	
Pinacyanol blue	Rh	2·2 mV	
Cu-Phthalocyanine	Rh	0·07 mV	
Violanthrene	Ag	1 mV	711
Violanthrene	Pt	2–3 mV	711
Violanthrene	Pb	4–14 mV	711
Chlorophyll	Pt	0·1 mV	712

of the current-voltage characteristics, the relation to intensity, etc.) suggest that the photovoltaic effect in the model arrangements which are imitations of sensitized systems, represents a kind of *pn*-photoeffect.

The fact must not however be overlooked that any discussion about a *pn*-photoeffect in semiconductor dye contacts presupposes a knowledge of various quantities about which very little is known, especially in the case of organic dyes. For example, no data are available regarding the lifetime τ, the length of diffusion or the concentration of the electrons and defect-electrons in unexposed systems. Whether any action, analogous to that in inorganic semiconductors is produced by donors or acceptors, and the combination of the latter with the *n*- or *p*-conduction of the dye is one of the questions which still remain to be answered.

Moreover, any theoretical treatment is rendered difficult because the *pn*-transfer is not of the familiar kind in which (as e.g. in germanium) the *n*- and the *p*-areas respectively have been produced by doping various regions in the same lattice in different ways. In the preparation of dye/semiconductor-contacts it is impossible to avoid the formation in the transition zone of additional lattice defects which, in the *pn*-zone, as in the case of two superimposed pieces of germanium of the *p*- and the *n*-type are manifested by increased recombination and therefore, amongst other things, by a decrease in the barrier action. It should also be noted that in semiconductor/dye systems contact is made between compounds having different lattice constructions. This influence is immediately revealed by the manifestation of a distinct rectifier action by a *pn*-layer which is built up of *p*- and *n*-conducting phthalocyanines, i.e. of organic semiconductors of the same constitution, whereas there is only a slight indication of it in inorganic/organic contacts.

In spite of this limitation it nevertheless seems appropriate to attribute the photovoltaic effects in inorganic/organic systems (and in pure organic *pn*-systems) to the influence of irradiation on the space charge regions between the semiconductors and the dyes (as a kind of *pn*-transfer), since this hypothesis

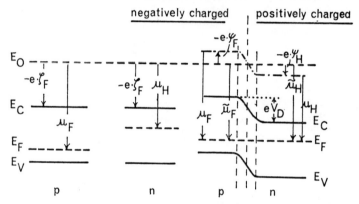

Fig. 57. Diagram of the energy bands of the *pn*-transfer.

conforms best with the previous experimental findings and also permits the behaviour of further inorganic/organic arrangements to be predicted.

Contact without Irradiation by Light: On the basis of this hypothesis the energy band diagram would correspond to the scheme represented in Fig. 57 when a semiconductor of the *n*-type is in contact with a dye of the *p*-type.

It will be assumed that the *p*-conduction in a dye is due to the influence of lattice defects which act as acceptors, as, say, when the dye is doped with *o*-chloranil, etc. To simplify matters the edges of the conduction and valency bands have moreover been drawn in the same position in the scheme, and the electrostatic potential of the semiconductor and dye, before making the addition, is established when $\psi = 0$, so that the Fermi energy E_F conforms with the chemical potential in both the compounds.

It can be seen from the diagram that when contact is made without the action of light an electrostatic difference in potential $\Delta\psi$ is built up between the inorganic semiconductor and the dye, and this leads to the formation of a space charge in the transition zone between the n- and p-regions. This upsetting of the equilibrium in the charge (i.e. in the neutrality) in the n- and p-regions is brought about in the boundary layer by the diffusion of electrons from the n- to the p-region owing to the fall in concentration (which is given by the gradients of the chemical potential) thus leaving behind a positive charge which is localized in the ionized donors, whereas defect-electrons diffuse from the p- into the n-region, leaving behind a negative charge which is localized in the acceptors. Equilibrium is established as soon as the current i_D which is formed by the diffusion of the charge carriers

$$i_D = -D \cdot e \cdot \frac{dn}{dx} \tag{78}$$

by which the field current i_F

$$i_F = -n \cdot e \cdot \mu_e \cdot \frac{d\psi}{dx} \tag{79}$$

flowing in the opposite direction and produced by the difference in the electrostatic potential of the n- and p-zones, is compensated, i.e. when

$$i_D = i_F = 0 \tag{80}$$

(In equation 78, $D = \mu_e \cdot kT/e$ is the diffusion coefficient and μ_e is the mobility.)

The difference in the electrostatic potential (macropotential) set up in the equilibrium between the dye and the semiconductor $d\psi = \psi_n - \psi_p$ corresponds to the so-called internal diffusion voltage V_D

$$V_D = \psi_n - \psi_p \tag{81}$$

which, since $i_D + i_F = 0$, bears the following relation to the concentrations of the charge carriers (electrons and positive holes, respectively) in the n- and p-regions:

$$\frac{dn}{n} = -\frac{\mu_e}{D} \cdot d\psi \tag{82}$$

$$n = N_o \cdot e^{-\frac{V_D \cdot e}{kT}} \tag{83}$$

or

$$V_D = \frac{kT}{e} \ln \frac{n \text{ (in the } n\text{-conductor)}}{n \text{ (in the } p\text{-conductor)}} \tag{84}$$

and

$$V_D = \frac{kT}{e} \ln \frac{p \text{ (in the } p\text{-conductor)}}{p \text{ (in the } n\text{-conductor)}} \tag{85}$$

From equations 84/85 it can be ascertained that the diffusion voltage V_D can be calculated if the concentrations of the electrons and the defect-electrons, respectively (n and p, respectively), which are present at a given time in a semiconductor and dye in contact are known.

Finally, this is due to the fact that the diffusion voltage V_D is formed as the result of the equalization of the variable chemical potentials μ of the systems in contact.[713] As Fig. 57 shows, the following holds:

$$e \cdot V_D = \mu_{F,p} - \mu_{H,n} \tag{86}$$

Since, moreover, the chemical potential of the electrons in general is given by

$$(E_F \equiv)\mu = E_c + kT \cdot \ln(n_0/N_c) \tag{87}$$

and on introducing the electron affinity $-e\zeta$, by

$$(E_F \equiv)\mu = -e \cdot \zeta + kT \ln(n_0/N_c) \tag{88}$$

in the state of equilibrium which is characterized by the identity of the electrochemical potentials ($\tilde{\mu}_F = \tilde{\mu}_H$) we have

$$-e\psi_F + \left[-e\zeta + kT \ln\left(\frac{n}{N_C}\right)\right]_F = -e\psi_H + \left[-e\zeta + kT \ln\left(\frac{n}{N_C}\right)\right]_H \tag{89}$$

$$\left[-e\zeta + kT \ln\left(\frac{n}{N_C}\right)\right]_F - \left[-e\zeta + kT \ln\left(\frac{n}{N_C}\right)\right]_H = (-e\psi_H) - (-e\psi_F) \tag{90}$$

Assuming for the sake of simplicity that $\zeta_F = \zeta_H$, it follows from equation 90 that

$$[-(kT/e) \ln(n/N_c)]_F - [-(kT/e) \ln(n/N_c)]_H = \psi_H - \psi_F \tag{91}$$

or, in conformity with equation 84:

$$V_D \equiv \psi_{H,n} - \psi_{F,p} = \frac{kT}{e} \ln \frac{n_H}{n_F} \tag{92}$$

Analogously, V_D can also be derived from the variable chemical potential of the positive holes of the dye and the semiconductor, since the following holds for the chemical potential (Fermi energy) of the positive holes:[713]

$$E_F = E_v - kT \ln(p_0/N_v) \tag{93}$$

$$E_F = -e \cdot \zeta - \Delta E - kT \ln(p_0/N_v) \tag{94}$$

The above discussion therefore appears to be worthy of note in connection with problems concerning inorganic/organic systems, because, above all things, it shows that in general when two compounds in which differences occur in the chemical potential of the electrons (positive holes)—i.e. for instance also between two variable *n*-conductors or *p*-conductors (!)—are placed in contact with each other, a space charge or an electrostatic difference in potential is set up between the two compounds. In theory it might not be of crucial importance whether the systems in contact belong to the inorganic semiconductors or to the "organic semiconductors" which likewise do indeed exhibit electronic conductivity.

It may however be assumed that a thermal equilibrium can be established, as for instance in the case of metal/semiconductor contacts or semiconductor/electrolyte contacts. For example, the transfer of electrons from the system (I) to a semiconductor of low conductivity (II) only takes place if the condition for the difference between the thermal work function W_{therm} of system I and the electron affinity of II $(-e\zeta)$ is fulfilled,

$$W_{\text{therm, I}} - e\zeta_{\text{II}} < 1\cdot 5 \ [\text{eV}], \tag{95}$$

or if positive holes in the valency band or deeply situated electron acceptors respectively are present in the semiconductor II.

Contact on Irradiation by Light: The photovoltaic effect observed in inorganic/organic and in pure organic model systems is explained in accordance with the theory of the *pn*-photoeffect by the fact that the diffusion voltage V_D which separates the pairs of electron/positive holes formed in the boundary layer on irradiation by light drives the electrons into the *n*-zone and the positive holes into the *p*-zone.[701,713–716] In the case of a short circuit this current continues as a short-circuit current I_0 in the outer circuit, and when the external resistance is very high it lowers the electrostatic potential in the *p*-zone and raises that in the *n*-zone, so that the potential gradient $\Delta\psi$ which is present in the unexposed state decreases. In this way a photoelectric voltage U_∞ is set up at the electrodes and in the limiting case this voltage can reach exactly the same value as the diffusion voltage.

According to the theories of the *pn*-photoeffect worked out by *Wiesner*, *Ruppel* and *Diehl*,[717,719] the short circuit current I_0 and the photovoltage U_∞ can be derived from equation 96, which is valid provided that $e^V \ll e^{V_D}$.[719]

$$I_{\text{phot}} = F \cdot e\left(e^{\frac{V}{v}} - 1\right)\left[\frac{n_p L_n}{\tau_n} + \frac{p_n L_p}{\tau_p}\right] - F \cdot e \cdot g(L_n + L_p) \tag{96}$$

where e is the elementary charge; $v = kT/e$; L_n, τ_n are the length of diffusion and the lifetime of the electrons; L_p, τ_p are the length of diffusion and the lifetime of the positive holes; n_p is the thermal electron concentration in the *p*-zone; p_n is the thermal defect electron concentration in the *n*-zone; g is the quota of the generation of carriers which is dependent on the region (charge carriers/cm³ . sec); and F is the area of the *pn* arrangement.

Short-circuit Current: Since V in equation 96 represents an external voltage which is applied to a *pn*-transfer, the following simplified expression is obtained in the case of the short circuit ($V = 0$):

$$I_0 = -F \cdot e \cdot g(L_n + L_p) \tag{97}$$

g gives the generation of carriers which is dependent on the region (charge carriers/cm^3 . sec) and which is produced by the absorption of radiation, so that the following relation is valid:

$$g = \eta \cdot g_0 \cdot e^{-\frac{d}{L_\lambda}} \tag{98}$$

where η is the quantum efficiency which was defined in Chap. V, section 2.5, and which states the ratio between the number of primarily excited charge carriers and the number of light quanta absorbed; L_λ is the depth of penetration of the light; d is the distance between the exposed surface and the *pn*-layer; g_0 is the number of quanta absorbed immediately below the surface which is connected with the number of photons incident per cm^2 and second (irradiation efficiency I_λ) by means of

$$I_\lambda = g_0 \cdot L_\lambda \tag{99}$$

Since in the experimental arrangements used here the depth of penetration of the light L_λ can be assumed to be great, and d is very small, equation 98 can be simplified by taking e^{-d/L_λ} as being approximately equal to 1. The short-circuit current is therefore given by:

$$I_0 = -eF \cdot \eta \cdot g_0 \cdot (L_n + L_p) \tag{100}$$

and by

$$I_0 = -eF \cdot \eta \cdot \frac{1}{L_\lambda} \cdot (L_n + L_p) \cdot I_\lambda \tag{101}$$

respectively.

This means that the short-circuit current increases in proportion to the efficiency of the irradiation I_λ, a state of affairs which was observed with dye/semiconductor arrangements (see equation 77). These equations show moreover that I_0 still depends on the lengths of diffusion of the electrons and positive holes L_n and L_p which are strongly influenced by traps.

Photovoltage: For the voltage $V = U_\infty$—cf. also[720]—which can be derived from equation 96 by inserting the term $I_{\text{phot}} = 0$ the following equation is valid:

$$U_\infty = \frac{kT}{e} \ln\left[1 + \frac{g(L_n + L_p)}{n_p L_n/\tau_n + p_n L_p/\tau_p}\right] \tag{102}$$

This signifies that, in conformity with the experimental findings, a logarithmic relation exists between the photovoltage of the *pn*-effect and the efficiency of the irradiation ($I_\lambda \approx g$) (see equation 76). It should also be noted that when the intensity of the illumination is high, U_∞ approaches the diffusion voltage V_D; cf. [717-719].

Since it is not yet possible to state accurately the values of L_p, τ_p, etc., of dye-photoconductors, no attempt to give a more detailed analysis will be made here. In the following a few of the experimental findings which argue in favour of the possibility of a *pn*-junction in inorganic/organic contacts will however be briefly discussed.

7.1.4 A SUMMARY OF THE FACTS WHICH ARE INDICATIVE OF A *pn*-TRANSFER IN MODEL SYSTEMS.

7.1.4.1 *Dependence on Direction.*
The dependence of the short-circuit current and the photovoltage on the direction of the *pn*-junction observed in a series of silver halide/dye and CdS/dye systems and of combinations of *n*- and *p*-conducting dyes conforms with the direction which can be expected from the theory of the *pn*-photoeffect, for *p*-conducting dyes become positively charged on exposure when they are combined with *n*-conducting semiconductors, whereas *n*-conducting dyes become negatively charged when they are combined with *p*-conducting photoconductors. An important fact is that only the *pn* theory can explain why, for instance, when Malachite green (*n*-conductor) is placed in contact with Merocyanine A 10 [59] (*p*-conductor) a passage of electrons from the Merocyanine to the Malachite green is measured on exposure.

It is therefore natural to examine from the point of view of the *pn*-hypothesis the photovoltaic effects in similar systems which have hitherto been published in the literature and for which special interpretations have been given from time to time.

a. A photovoltaic effect was observed by *Grossweiner et al.*[420,721] in Au/ZnO/Eosine (Na)/Au systems (distance between the electrodes 1 cm), which effect indicates that electrons are transferred from the dye to the ZnO during exposure. An argument for the formation of the *pn*-photoeffect is firstly that ZnO is an *n*-type photoconductor—cf. e.g. *Mollwo et al.*[722]—and Eosine (Na) is a *p*-conducting dye.[53] Secondly, the thermal work function of ZnO ($\approx E_{F,\text{ZnO}}$) lies at 4·84 eV[251] and we already know that the corresponding value for Eosine (Na) ($\approx E_{F,\text{Eosine}}$), is between 5·0 and 5·5 eV[408,688]. This signifies that the conditions might be given for the setting up of a difference in potential between ZnO and Eosine in the unexposed state and therefore for the *pn*-photoeffect.

On the basis of more recent results with ZnO/dye systems, *Grossweiner* and his collaborators[723] have come to the conclusion that the photovoltaic effect in these systems is due to the setting up of a *Schottky* boundary layer potential between the ZnO and the dye in the dark, and this potential brings about a transfer of electrons from the dye to the ZnO on exposure.

b. Aromatic amines which are graded as *p*-conductors likewise give photovoltaic effects on being coupled with *n*-conducting dyes—e.g. Crystal violet.[724] The observed directions of the photocurrents are in conformity with the *pn* theory.

c. In systems consisting of layers of TlI (*p*-type) and *n*-conducting dyes— e.g. Phenosafranine—the dye becomes negatively charged on exposure and the inorganic semiconductor positively charged.[404] This is also in harmony with the above theory.

Table 22 also shows the agreement which exists in a number of systems between the directions of the photocurrents and photovoltages respectively arrived at from observation and by derivation from the *pn* theory.

TABLE 22

Photoconductor		Dye		Charge on exposure				Literature
				Exp.		pn-theory		
I		II		I	II	I	II	
CdS	(*n*)	Phthalocyanine	(*p*)	−	+	−	+	
CdSe	(*n*)	Merocyanine A 10 [59]	(*p*)	−	+	−	+	
ZnO	(*n*)	Eosine (Na)	(*p*)	−	+	−	+	420, 721, 723
AgBr	(*p*)	Rhodamine B	(*n*)	+	−	+	−	
AgI	(*p*)	Malachite green	(*n*)	+	−	+	−	
TlI	(*p*)	Phenosafranine	(*n*)	+	−	+	−	404
Malachite green	(*n*)	Merocyanine A 10	(*p*)	−	+	−	+	
Rhodamine B	(*n*)	Merocyanine A 10	(*p*)	−	+	−	+	
Aromatic amines	(*p*)	Crystal violet	(*n*)	+	−	+	−	724
Mg-Phthalocyanine	(*p*)	N.N'.N'.-Tetramethyl-*p*-phenylenediamine		+	−	(+	−)	682

The observation that photovoltaic effects likewise occur when an *n*-type photoconductor is combined with *n*-conducting dyes must not be taken as being contradictory to the *pn* theory,[725] for, according to equation 86, the difference in the chemical potentials of the electrons (or positive holes) in the compounds in contact is a decisive factor in the formation of the diffusion voltage V_D (which corresponds to the difference in the electrostatic potentials and which finally produces the photovoltaic effect). According to equations 84 or 85 (and equation 92 respectively) the occurrence of a diffusion voltage and of a small photovoltaic effect must be reckoned with when there are only small differences in the electron or hole concentrations of compounds of the same type of conduction.

The examples given in Figs. 58 and 59 show this *nn'*-photoeffect in a CdS/CdSe cell which was produced in accordance with the arrangement I (Fig. 50), and which exhibits photovoltages of up to a maximum of 30 mV and short-circuit currents of the order of magnitude of 10^{-10} A. CdS and CdSe both represent n-conductors.

Since the concentration of the electrons in n-conducting dyes can certainly be taken as being low in comparison with that in CdS, it can be understood why electrons are transferred from the dye to the CdS, e.g. in CdS/Malachite green (n) or CdS/Rhodamine B (n) arrangements when the latter are exposed to light. This is because the concentration gradients of the electrons present

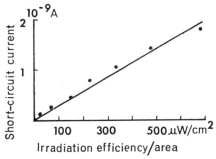

Fig. 58. CdS/CdSe. The short-circuit current as a function of the irradiation efficiency ($\lambda = 5870$ Å).

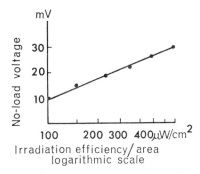

Fig. 59. CdS/CdSe. The photovoltage as a function of the irradiation efficiency ($\lambda \times 5870$ Å).

in the dark cause a positive space charge to build up in the unexposed boundary layer in the CdS zone and a negative one in the dye zone. The effects which are observed when a p-conducting semiconductor is placed in contact with dyes of the p-type can be explained in an analogous way. Corresponding with the mechanism discussed here, the setting up of a space charge between unexposed systems when placed in contact in most of the organic photoelements described in the literature might also be the cause of the photoelectric voltages and short-circuit currents measured. As examples we would mention the arrangement which was patented by *Calvin* and *Kearns*[726] and which consisted of magnesium pthalocyanine and tetramethyl-*p*-phenylenediamine (maximum efficiency $N_{\max} = 3 \cdot 10^{-12}$ W; $E_\infty = 200$ mV); also systems of thin layers of phthalein and triphenylmethane dyes ($N_{\max} = 6 \cdot 10^{-10}$ W; $I_0 = 0 \cdot 5 \cdot 10^{-9}$ A/cm²; $E_\infty = 100$ mV),[727] and cells of *n*-conducting poly-copper-phthalocyanine and *p*-conducting edge layers[728] or of dye-sensitized cationic and anionic exchange membranes.[729]

7.1.4.2 Dependence on Intensity. Equations 101 and 102 show that the linear and logarithmic relation between the short-circuit current and the photovoltage, respectively, and the efficiency of the irradiation, measured in the model inorganic/organic systems, can be explained by means of the theory of the *pn*-photoeffect. From the series of measurements shown in Figs. 58 and 59 it can, moreover, be seen that an analogous relation also holds for *nn'* systems.

7.1.4.3 Dependence on the Spectrum. According to the theory of the *pn*-photoeffect, the photoelectric short-circuit current and the photovoltage are due to the charge carriers formed in the boundary layer on absorption of light. It can therefore be understood why a photovoltaic effect is set up when two photoconductors with different absorption bands are placed in contact and irradiated in the region of absorption of one or other of the components.

In conformity with this is the fact that the action spectrum of the photovoltaic effect in semiconductor/dye arrangements exhibits sensitivity in the spectral region of the absorption of the inorganic semiconductor as well as in the region of absorption of the dye: cf. Figs. 55, 61, 65.

7.1.4.4 Current-voltage Characteristics. The lack of symmetry in the current voltage characteristics of exposed inorganic/organic and pure organic *pn*-systems is, as has already been discussed above (see section 1.2.6) in harmony with the effects which can be expected from the *pn* theory. Moreover, the arrangements prepared from *p*- and *n*-conducting layers of phthalocyanine show a rectifier action in the unexposed state as well, a fact which is in conformity with the *pn* theory.

When discussing these effects it must be borne in mind that the method of preparation used here can give rise to the formation of numerous lattice defects which promote the recombination of electrons and positive holes and which can therefore diminish the barrier action of the arrangements.[717]

7.1.4.5 The Influence of Doping. The photovoltaic effect in inorganic/organic systems can be enhanced by doping *p*-conducting dye components with electron acceptors, as illustrated in Fig. 60, which gives a CdS/merocyanine cell as an example.

If the *pn* mechanism discussed here for model inorganic/organic systems is valid, then the effect of doping will readily be understood; for, according to equation 104, the increase produced in the density of the positive holes p in equilibrium, by doping, will decrease the density of the electrons n, so that according to equation 84, the diffusion voltage V_D in the boundary layer between the *n*-conducting CdS and the *p*-conducting dye, and therefore, according to this observation, the photovoltaic effect, should increase.

$$n . p = \text{const} \tag{103}$$

and
$$n . p = n_i^2 \tag{104}$$

respectively.

Here n_i denotes the intrinsic conduction density ($n = p = n_i$) for which, e.g. for germanium, the following is valid:

$$n_i = 2\cdot 5 \ . \ 10^{13} \text{ cm}^{-3}$$

Even if there are no data of the n_i of dyes, and moreover the question of the action of the added acceptor molecules (which are considered here to be entirely analogous to inorganic semiconductors) is problematical,[53] nevertheless

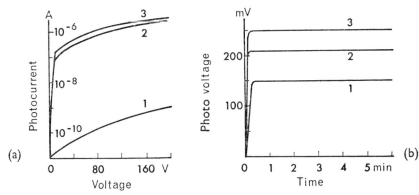

Fig. 60. The influence of doping on the photoconductivity (a) and the photovoltaic effect (b) in a CdS/Merocyanine A10 cell (Arrangement I of Fig. 50). (1) Undoped Merocyanine A10 [59] (2) (1) + tetraiodophthalic anhydride (10^{-7} mol/cm²). (3) (1) + iodine (10^{-8} mol/cm²). Note: The statement mol/cm² signifies that under the same conditions in each case x mols were brought on to a screen cell of 1 cm² in area.

a comparison between the orders of magnitude of the experimental values obtained for e.g. CdS/merocyanine cells and those of the data which can be derived from other semiconductors, does appear to be possible.

When CdS ($n_n \approx 2\cdot 5 \ . \ 10^{16}$ cm^{-3}) is combined with an undoped dye (assuming that $n_p \approx 2\cdot 5 \ . \ 10^{13}$ cm^{-3}), according to equation 84 a value of about 180 mV would be obtained for the diffusion voltage V_D. Assuming that the density of the acceptor and therefore also that of the positive holes is increased to $2\cdot 5 \ . \ 10^{15}$ cm^{-3} by doping the dye, e.g. with *o*-chloranil (see Chap. VI, section 2.2.5.2), according to equation 104 (assuming that $n_i = 2\cdot 5 \ . \ 10^{13}$ cm^{-3}) a density of the minority carriers $n_p = 2\cdot 5 \ . \ 10^{11}$ cm^{-3} and therefore a diffusion voltage of V_D of about 300 mV could be set up. The experimental increase in the photovoltage up to an order of magnitude of 300 mV can therefore be explained as being the result of doping the dye components.

In what way can the results obtained with model systems have any significance in relation to the mechanism of spectral sensitization?

7.2 Application of the Results Obtained with Model Arrangements of Sensitized Systems to the Mechanism of Spectral Sensitization

7.2.1 THE *pn* THEORY OF SPECTRAL SENSITIZATION. The experiments suggested regarding the spectral sensitization effect as a kind of *pn*-photoeffect,[416,470] for it can be seen from the investigations that the absorption of light by the dye not only enables photoconductivity to be developed in semiconductors which absorb in the short wavelengths and which are in contact with the dye, and in the case of silver halide, the elementary photographic process to be released, but it also enables photovoltaic effects such as those which are known to occur in inorganic *pn*-photoelements to develop in semiconductor/dye contacts. Since various regularities in the photovoltaic *pn*-effect in semiconductor/dye systems are also characteristic of spectral sensitization, it appears possible to discuss the spectral sensitization of the photoconductivity and of the primary photographic process (at least in the case of multimolecular layers of sensitizer) on the basis of a mechanism which is valid for both effects. According to this hypothesis it can be assumed that an internal electric field is set up in the dark between the semiconductor and the sensitizer as the result of a transfer of electrons (transfer of positive holes) from the region of higher concentration of charge carriers to that of lower concentration, and this electric field separates the charge carriers formed in the light. Finally, the difference between the chemical potential of the electrons or positive holes in the semiconductor and in the dye, as well as a general tendency for these potentials to equalize, is responsible for the development of an internal field in the unexposed semiconductor/dye contact and therefore for the transfer of electrons from the exposed dye to the semiconductor and vice versa—i.e. for the spectral sensitization.

Assuming that the dye and the semiconductor are in close contact and also that equalization of the charge carriers is physically possible, the conditions for the spectral dye sensitization of a photochemical reaction and of a sensitized photoelectric effect, respectively, in semiconductors can be established by the general conditions set out below if the hypothesis discussed here is valid:

$$\text{Sensitization of an } n\text{-conductor} \quad |\mu_H| - |\mu_F| < 0 \quad (105)$$

$$\text{Sensitization of a } p\text{-conductor} \quad |\mu_H| - |\mu_F| > 0 \quad (106)$$

$$\text{Photographic sensitization} \quad |\mu_{\text{AgBr}}| - |\mu_F| < 0 \quad (107)$$

$$\text{Photographic desensitization} \quad |\mu_{\text{AgBr}}| - |\mu_F| > 0 \quad (108)$$

Where μ_H, etc., denote the chemical potentials of the electrons in unexposed semiconductors and dyes.

It can be deduced from the relations 105–108 that dyes of the *p*-type (*n*-type) might be most suitable for sensitizing an *n*-type semiconductor (*p*-type semiconductor) since in these cases the possibility of a difference in the chemical potentials of the systems in contact is particularly obvious. However, the examples also show that an *n*-conducting system can by all means be sensitized by dyes of the *n*-type, because differences in the chemical potentials of the electrons in these systems cannot be excluded. Similar considerations also apply to *p*-conducting substrata.

The relations 105–108 do of course represent the fundamental process of sensitization in a greatly simplified form. However, in the event of the stated mechanism of sensitization being valid, recourse could possibly also be had to these conditions in order to predict the sensitizing possibilities of any dyes which are capable of being adsorbed to various photoconductors. Since, moreover, the chemical potentials of various dyes and the relation between these chemical potentials and the constitution of the dyes have to be known, it was deemed necessary above all to determine these quantities accurately by experiment.

7.2.2 EXPERIMENTAL INDICATIONS OF THE VALIDITY OF THE *pn* THEORY OF SPECTRAL SENSITIZATION. The *pn*-mechanism which has been derived from the photovoltaic effect in model systems is able to explain most of the effects which are characteristic of the spectral sensitization reaction (see Chap. III), as is shown by the following discussion.

7.2.2.1 *Relation to the Spectrum*. It can be seen from Fig. 61, for example, that the spectral sensitivity of the photoconductivity of cadmium sulphide which has been sensitized by a layer of merocyanine (Merocyanine A 10) corresponds with the relation between the photovoltaic effect and the wavelength in the model CdS/Merocyanine A 10 system. Since the sensitized photoconductivity of CdS was measured in a sandwich type cell which largely corresponds with the model CdS/Merocyanine arrangement, this conformity may be regarded as an indication that a relation exists between the formation of a photovoltage by semiconductor/dye contacts on the one hand and the spectrally sensitized photoconductivity on the other hand.

Analogous observations made with silver halide/dye systems show, moreover, that spectral photographic sensitization is also in harmony with the *pn* theory.

7.2.2.2 *Relation to Direction*. The directions of the photovoltage and short-circuit currents observed in model systems prepared from semiconductors and dyes may be taken as experimental proof of the exchange of electronic charge carriers in a semiconductor/dye contact. This *pn*-photoeffect theory makes it clear why not only the photoconductivity of an *n*-conductor but also that of a *p*-conductor can be increased when the dye is exposed, for which observation neither the resonance hypothesis nor the simple electron transfer theory of *Gurney* and *Mott* is able to furnish an explanation. Whereas, according to the diagram in Fig. 62, electrons are driven from the dye to the semiconductor by the internal electric field when a *p*-conductor is sensitized, the flow of electrons in the reverse direction, which is equivalent to an increase in the density

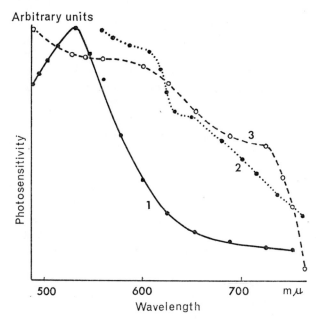

ig. 61. The spectral distribution of the photosensitivity for an irradiation efficiency of $10^3 \mu W/cm^2$ (1) CdS-layer. (2) CdS, sensitized by Merocyanine A10 (thickness of layer 0·1 μ), (photoconductivity; measurement in a sandwich type cell). (3) CdS/Merocyanine A10-system. (Photovoltaic effect, measured in arrangement II of Fig. 50.)

of the defect electrons in the *p*-conductor, likewise follows directly from the *pn* theory.

The spectral sensitization of *p*- and *n*-conductors therefore becomes understandable.

The results obtained with Ag/silver halide/dye/Ag arrangements—cf. for example Fig. 63—reveal that the said hypothesis is also able to describe the spectral sensitization of the photographic process. For, whereas according to the *pn* theory, in combinations of AgBr and AgI respectively with dyes of the *n*-type (Malachite green, etc.), the silver halide becomes positively charged and the dye negatively charged when the dye is irradiated in its absorption band, in arrangements consisting of silver halides and *p*-conducting sensitizing dyes, the direction of the photocurrent and the photovoltage is reversed: in each case the dye becomes positively, and the silver halide negatively, charged. This signifies that in systems which are imitations of sensitized photographic emulsion layers it has been established that electrons are transferred from the dye to the silver halide, which is namely the prerequisite for the formation of the latent image.

This suggested attributing the direction of the transfer of electrons which is dependent on the nature of the adsorbed dye to differences in the chemical

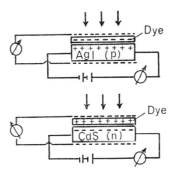

Fig. 62. Diagrammatic representation of the connection between the photovoltaic effect in semiconductor/dye systems and the spectral sensitization reaction.

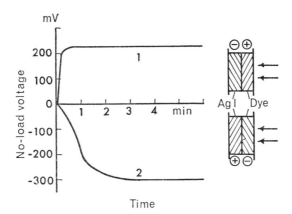

Fig. 63. The photovoltage as a function of the time of exposure. (1) AgI/Merocyanine FX 79 [60] (p-type). (2) AgI/Rhodamine B (n-type). Recorded curves, redrawn. Arrangement I of Fig. 50. $\lambda = 6000$ Å.

potentials of the dye components (equations 107 and 108). The fact that this variable behaviour also has an effect on the type of conduction in the dyes in the examples given here would be in harmony with this assumption.

It should also be noted that the resonance theory of sensitization is unable to provide any explanation of these observations.

7.2.2.3 *Desensitization.* The transfer of electrons from the silver halide to the dye which is observed on exposure of a series of silver halide/dye systems such as that which occurs for example when *n*-conducting Malachite green is built into a model photovoltaic semiconductor/dye system, can possibly be brought into relation with the desensitizing action of the dye components. According to the boundary layer hypothesis the electric field set up between the dye and the silver halide would displace the electrons formed in the light towards the dye, but any positive holes which might be formed would be displaced towards the silver halide, thus diminishing the photographic action. The desensitization effect which is often simultaneously combined with dye-sensitization (see Chap. I) cannot of course be explained in an analogous manner. This therefore suggested that in this case the more or less large decrease in the natural sensitivity of the silver halide was attributable to a partial irreversibility of the electron transfer processes which lead to the setting up of the internal electric field. In addition, the proportionality between the number of electrons which has been transferred to the adsorbed dye and the number of charge carriers in an unsensitized substratum (ZnO), as shown experimentally by *Grossweiner et al.*,[723] indicates that a portion of the electrons which are formed in the substratum itself on exposure, can also be removed from the semiconductor. The natural sensitivity of exposed sensitized systems may therefore be less than that of the unsensitized substrata.

Further experiments are desirable in order to elucidate this question.

It is worth noting that the negative charging up of a dye which is characteristic of desensitization could also be measured in silver halide/pinacyanol arrangements (i.e. in a silver halide/sensitizer arrangement). The direction of the charge conforms with the state of affairs discussed in Chapter I, according to which fairly thick layers of typical sensitizing dyes are able to assume preponderantly desensitizing properties. This may possibly be attributable to the fact that the increasing influence of the anions with vacant sites (I^-, etc.) in the dye with increase in layer thickness has somewhat the effect of additional built-in donors which raise the concentration of the electrons and therefore the chemical potential of the dye. This assumption would conform with *Nelson*'s statement (see Chap. VI, section 2.2.1) to the effect that the sensitizing power of pinacyanol layers can be decreased or removed (desensitization) by improving their dark conductance.

In this connection mention should also be made of the correlation discussed in Chap. VI, section 2.2.6 between the photographic properties of dyes of similar constitution on the one hand and their conductance and absorption behaviour on the other hand, for the tendency of dyes of similar constitution to develop *n*-conduction which, in desensitizers, runs parallel with their desensitizing action, likewise indicates a difference in the chemical potentials of sensitizing and desensitizing dyes.

7.2.2.4 The Repeated Reactivity of a Dye. According to sections 1.2.2 and 2.3.2 of Chaps. III and IV, respectively, a dye molecule can take part several times in a sensitization process, so that a process of regeneration (balance of charge) in the dye during sensitization has to be reckoned with. This regeneration of the dye is also indicated by the photovoltaic effect in dye/semiconductor systems, as otherwise destruction of the dye layer would make any lengthy measurements impossible. The analogy between the electronic processes taking place in these model systems and the sensitization reaction makes it probable that the dye is regenerated in the same manner in both cases.

According to the theory of the *pn*-photoeffect, the photovoltage is due to a decrease which occurs in the potential gradient in the boundary layer and which is brought about by the charge carriers formed on exposure. A steady photovoltage is set up because the lowering in the electrostatic potential difference must be balanced by an internal reverse current, which, e.g. in CdS/dye (*p*) cells represents a transfer of electrons from the CdS to the dye. Since the spectral sensitization reaction corresponds as it were with the case of a photovoltage (open or missing terminals), it can be assumed that the same electronic processes take place in photoconductors which are covered with multimolecular layers of dye. In this connection it should be noted that in practical photography the ability of the dye molecules to react several times is not necessarily a prerequisite for the production of a latent image, since the 10^5 to 10^6 dye molecules which are adsorbed to a silver halide grain only have to transmit from 4 to 500 absorbed quanta to a grain.

7.2.2.5 Supersentization. The fact that it is possible to influence the photovoltaic effect in dye/semiconductor systems by doping the dye components, as discussed in section 2.2.5 of Chapter VI, is apparently closely connected with the effect of supersensitization, for it is natural to consider that the increase in the sensitized photoconductivity caused by the doping agent (cf. also *Inoue et al.*[452]) is a result of the change in the concentration of the charge carriers in the sensitizer due to the added electron acceptors.

According to the *pn* theory the action of a supersensitizer would consist primarily in a decrease in the concentration of the electrons in the layer of sensitizer in accordance with equation 104, and therefore in the chemical potential of the electrons in the dye. When in contact with a suitable semiconductor this could lead to an increase in the difference in the electrostatic potentials between the semiconductor and the dye, and therefore to an increase in the sensitization, depending on the concentration and the nature of the electron acceptor.

The example reproduced in Fig. 64 of the influencing of the photovoltaic effect in a silver halide/dye arrangement by doping indicates that the mechanism under discussion also describes the supersensitization of the primary photographic process. Fig. 65 shows, moreover, that no change is brought about in the spectral sensitivity in the long wavelength region by the traces ($< 10^{-7}$ mole per cm² of cell area) of supersensitizer added to the silver halide/dye.

7.2.2.6 *Other Effects.* From the examples given it is already evident that the *pn* theory can explain various regularities in the mechanism of spectral sensitization, and therefore no attempt will be made here to enter into a discussion of other effects—the influence of temperature (given by equation 102),

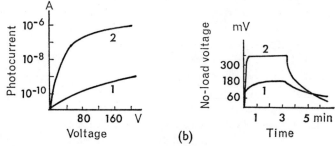

Fig. 64. The influence of doping on (a) the photoconductivity and (b) the photovoltaic effect in AgI/dye cells. (a) Photoconductivity: (1) Merocyanine A10 (59). (2) Merocyanine A10 + o-chloranil (10^{-7} mol/cm²; see note under Fig. 60). (b) Photovoltaic effect: (1) AgI/Merocyanine A10. (2) AgI/Merocyanine A10 + o-chloranil. Recorded curve, redrawn.

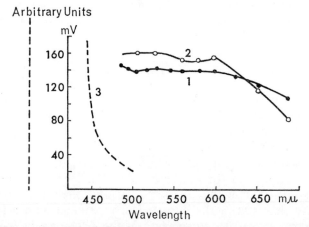

Fig. 65. (1) AgI/Merocyanine FX 79 (Formula, see Table 8). Spectral distribution of the no-load voltage for an irradiation efficiency/area of 10^3 μW/cm². (Arrangement I of Fig. 50.) (2) (1) + p-chloranil (10^{-7} mol/cm²; see note under Fig. 60). (3) AgI. Spectral distribution of the photosensitivity. After Terenin, et al.[402]

the decrease in the sensitizing action e.g. of haematoporphyrine[404] on being converted to the reduced form (due to a decrease in μ), etc.—which are likewise in harmony with the stated theory.

In the following only a brief account will be given of the extent to which the mechanism of sensitization which has been derived from studies of multi-molecular layers of dye is also applicable to monomolecular layers.

7.2.3 THE PROBLEM OF SENSITIZATION WITH MONOMOLECULAR LAYERS OF DYE. The electron transfer theory of spectral sensitization derived from studies of model inorganic/organic *pn* arrangements attributes the sensitization reaction to two effects: firstly, to the formation and migration of free charge carriers in the sensitizer by the absorption of radiation, and secondly, to the formation of a space charge in the boundary zone between the dye and the semiconductor.

The problem discussed here therefore requires an answer to the question as to whether these assumptions are fulfilled when semiconductors are brought into contact with monomolecular layers of dye.

7.2.3.1 *The Photoconductivity of Monomolecular Layers of Dye.* The quantum yield of the photoconductivity increases—cf. Chap. V, section 3.5—with decrease in layer thickness (of the order of magnitude of from 10^{-4} to 10^{-6} cm). This state of affairs and the observation of photoconduction in a monomolecular layer of polythienylene built into barium arachidate[626] prove that electronic charge carriers which are capable of moving about under the influence of an electric field, are also formed in monomolecular layers of dyes on irradiation in the absorption band of the dye.

This, therefore, gives the first prerequisite for the above mechanism of sensitization.

7.2.3.2 *The Formation of a Space Charge between a Dye and a Semiconductor. The Manner in which the Dye is Deposited on the Layer:* Previous knowledge of the subject indicates that in general one has to reckon with dyes deposited in the form of fairly large, aggregated units on the semiconductor grains (Fig. 16). In the case of a silver halide grain of $1\ \mu^2$ in size these units may consist of up to a maximum of 10^5–10^6 dye molecules deposited on top of one another, whereas with a lesser degree of coverage deposits of smaller, island-shaped units of 10^3 molecules might be present.

In this connection the reader is reminded of a fact which is known from the assimilation of CO_2, namely that "photosynthetic units" consist of about 2500 molecules of chlorophyll.[53,731–734]

The Chemical Potential of the Sensitizing Units: Indications as to the nature of the conduction (*n*- or *p*-type) of these dye units undoubtedly appear to be problematical. This is because, firstly, no statements can be made regarding the donor or acceptor actions of anions or cations containing vacant sites in such units consisting of basic or acid dyes. Secondly, the observations made by *Terenin et al.*[735] show that the type of conduction in a dye film can be altered by influencing the state of aggregation (crystalline–amorphous). Moreover, the action of the grain boundaries[736,737] which act as acceptors, the influence of the surrounding atmosphere of gas and, in a photographic emulsion, the presence of additional foreign molecules (gelatin, etc.) must also be taken into consideration.

In spite of the lack of data on the type of conduction in dye units which are deposited in close contact with the surface of a grain (condition of sensitization 2, see Chap. III, section 1.1) and which in many cases consist of flat dye molecules deposited parallel to one another (condition of sensitization 1, see Chap. III, section 1.1), it might however in theory be impossible to exclude the presence of a low concentration of charge carriers which are formed thermally by intrinsic excitation and from lattice defects. Moreover, the electrons occur in the "conduction band" which has arisen from the first excitation singlet state, whereas the gaps in the "valency band" can be considered to be positive holes; cf. also *Szent-Györgyi*.[476] According to this assumption, chemical potentials, for which the general relations[88,94,738,739] are valid, can also be ascribed to the electrons (positive holes) in the dye units. The electron affinity $-e\zeta$ and the distance between the energy bands ΔE is therefore apparently in agreement with the values which are known to be possessed by thick layers of dye and by crystals, for firstly according to *Scheibe's* rule the following is universally valid for dyes

$$-e\zeta \approx 3\cdot 4 \text{ eV} \qquad (109)$$

and secondly the sensitization spectrum in the case of monomolecular layers of dye is largely identical with the absorption spectrum of multimolecular layers (see Chap. III, section 1.2.7).

Establishment of the Equilibrium: As in the case of the model semiconductor/dye arrangements, the magnitude of the chemical potential of the electrons in adsorbed dye units might decide whether an exchange of electrons will take place, and therefore whether an electric field will be set up between the dye and the semiconductor. For equilibrium to be established the chemical potentials of the semiconductor and of the dye unit must be adjusted by a transfer of electrons in order to charge up the semiconductor positively or negatively depending on the mutual situation of the chemical potentials.

The establishment of the equilibrium and therefore the formation of the space charge layer presupposes, however, that for example in the case in which $|\mu_H| < |\mu_F|$ it is possible by satisfying equation 95, and by means of the presence of gaps in the fundamental band of the dye, respectively, for electrons to be transferred from the semiconductor to the dye. Single molecules which are characterized by an excitation level which is situated above the conduction band of the semiconductor, and by an occupied ground state, are unable to take up any electrons from the semiconductor. The prerequisites for the taking up of electrons might however be given in dye units in which, under the influence of lattice defects, which also include a possible acceptor action on the part of the intercalated oxygen atoms,[740,741] holes may be present in the "fundamental band."

This signifies that, amongst other things, a potential difference $\Delta\psi$ can be set up between the dye and the semiconductor in adsorbed monomolecular dye aggregates under the influence of which electrons which are formed in the dye on irradiation by light are transferred to the semiconductor. When specific electron acceptors are intercalated it is also necessary to reckon with an increase in the concentration of positive holes, and, as in the case of supersensitization, in that of the multimolecular layers of dye (see section 2.2.5), with an increase in the sensitizing action.

In this connection it should be pointed out that it appears questionable as to how far a single dye molecule can be considered to be capable of releasing a sensitization reaction on its own; for the power of dyes to react several times has not been established for single molecules, but has been derived from

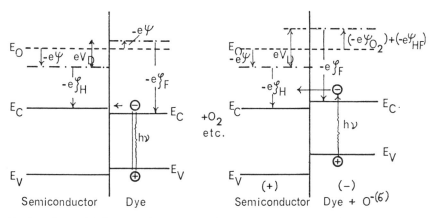

Fig. 66. Greatly simplified energy diagram in explanation of the influence of O_2 on a conductor/dye system.

measurements of the photolysis of silver bromide sensitized by layers of dyes, or from the sensitized photoconductivity of semiconductors. In practice, in most cases it might be that the deposits which consist of units of several dye molecules only sensitize in the mass, in analogy say to the process occurring in photosynthesis, so that the conditions for the regeneration and repeated transfer of electrons by the dyes are laid down from the outset.

Barrier Layer Semiconductor/Dye/Oxygen. It is also natural to ascribe the electrostatic potential difference which is set up between the dye and the semiconductor during the sensitization reactions which take place in the air, first and foremost to the anionic chemisorption of oxygen. This is because oxygen, by taking up one electron from the dye layer, can be adsorbed as $O^{-(\sigma)}$—cf. for example [564,569,742]—thus raising the electrostatic potential of the dye unit. The diagram in Fig. 66 illustrates this formation of (or increase in) the diffusion potential V_D which is due to the chemisorption of $O^{-(\sigma)}$ and under the influence of which the electron/hole pairs which have formed in the dye during exposure, become separated.

A simplified representation was chosen because it shows clearly not only the formation of the electrostatic potential difference ($=V_D$) but also the raising of the excitation level of the dye and the position of this level in relation to the conduction band of the semiconductor, which has a decisive influence on the process of photographic sensitization, in the unexposed state. For instance, it can be ascertained from the diagram that by the chemisorption of electron acceptors which are stronger than O_2 the excitation state of desensitizing dyes may possibly be raised to the level of the conduction band of the semiconductor owing to an intensified increase in the electrostatic potentials. This conforms with observations[251] that the condenser-photoeffect in thallium halides which

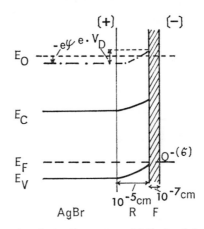

Fig. 67. Energy diagram of a semiconductor/dye system with the bands bending upwards at the surface.

have been sensitized by Malachite green (desensitizer) is spectrally sensitized by the action of bromine.

It must be noted that in contrast to the electrochemical potential $\tilde{\mu}$, neither the electrostatic potential $(-e\psi)$ nor the chemical potential μ in systems which are placed in contact can be regarded as constant. The reduction in the number of electrons which are transferred to the dye and to the chemisorbed $O^{-(\sigma)}$ respectively in the semiconductor therefore causes the bands to bend upwards in the manner shown in Fig. 67.

The importance of the surface barrier layer, which is built up by oxygen and which may possibly superimpose itself on the internal potential gradient which is already present *in vacuo* between the semiconductor and the dye units, lies in the fact that the electrons which are formed in the dye layer by the action of light are displaced towards the semiconductor, and therefore when the excitation level of the dye is in a suitable position, electrons can pass into the conduction band of the semiconductor. The positive holes on the other hand are attracted by the $O^{-(\sigma)}$ adsorption layer and discharged by the latter. The dye thus returns to its original neutral state and is therefore again able to form electron/positive hole pairs for the sensitization reaction. (The condition of repeated reactivity). Since oxygen becomes chemisorbed afresh, then when the exposure is fairly long a state of equilibrium is established by means of an internal reverse current which tries to compensate the difference in potential which was lowered on exposure.

It can be expected that the situation in the surface barrier layer will become fairly complicated when gases and vapours with different donor or acceptor actions are chemisorbed. Even when multimolecular layers of dye are used instead, the influence of the space charge layer which is due to the chemisorption of gaseous oxygen can be destroyed by the *pn*-junction which is built up between the dye and the semiconductor, an effect which may possibly be the

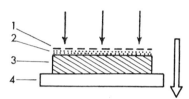

Fig. 68. Diagram of the measuring cells used for studying the influence of gases. (1) Semitransparent electrode (Rh, etc.). (2) Chemisorbed layer of gas (O_2, H_2). (3) Layer of dye (precipitated from solution). Exposure area 1 cm². (4) Electrode.

cause of the desensitizing action of multimolecular layers of dye (cf. Chap. III, section 2.1).

7.2.3.3 *Experimental Indications.* Finally, a brief mention will be made of a few observations which are in harmony with the hypothesis given for sensitization by monomolecular layers of dye.

1. *Putzeiko* and *Terenin*[743] have established that the dark conductivity of CdO, ZnO and other semiconductors decreases on adsorption of Malachite green and Methylene blue—which sensitize the photoconductivity of these semiconductors—as is to be expected if the relation discussed above is valid. The authors explain this by assuming that, by taking up a conduction electron from the semiconductor, the dye cations are adsorbed neutrally, whereas the negative counter ions in the *n*-conductor produce a surface barrier layer which has become impoverished in electrons owing to chemisorption. Even if this mechanism cannot be agreed to here, since indeed non-ionic dyes likewise sensitize, this hypothesis does nevertheless indicate the possibility of the formation of a space charge layer and of the influence of the latter on the electron transfer reaction of sensitization discussed here.

2. In experimental cells of the kind shown in Fig. 68, in which the

semitransparent electrode used for testing the possible influence of a gas is placed under only slight pressure on the dye layer, whereas the other electrode is firmly connected to the dye, a relation was found to exist between the short-circuit currents and the photovoltages (of the order of magnitude of 10^{-8} to 10^{-10} A and 0·1–2 mV, respectively, when the efficiency of irradiation was $10^3\,\mu\mathrm{W/cm^2}$), owing to adsorbed gases.[497,644]

The reversibility of the action of gases observed with air and oxygen respectively (Fig. 69), the electron rectification as well as the barrier effects support

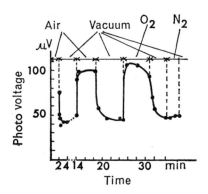

Fig. 69. Phloxine (arrangement of the cells as in Fig. 68). The photovoltage as a function of the action of gases (air, O_2) with time. Irradiation efficiency $10^3\,\mu\mathrm{W/cm^2}$. Intermittent exposure 50 Hz.

the assumption discussed in section 2.3.2 to the effect that a surface barrier layer is built up by the anionic chemisorption of oxygen to the dye, and that a photocurrent can be formed by the influence of this barrier layer.

In conformity with this hypothesis, hydrogen, for which cationic chemisorption

$$H_2^{(g)} \rightleftharpoons 2H^{+(\sigma)} + 2\ominus^{(R)}$$

where (g) is the gas phase, (σ) is the chemisorption or the σ-phase and (R) is the barrier layer near the surface.

with formation of a positively charged surface layer is a controversial point, causes the electrons released in the dye on exposure to flow in the reverse direction. Cf. the series of measurements which are given as an example in Fig. 70, and further details (e.g. the influence of metal electrodes, etc.) in.[53,507]

The measurements might be regarded as a proof that surface barrier layers are formed in dye films by the chemisorption of gases and vapours (electron acceptors, donors) and that these surface barrier layers can bring about a separation of the electron/hole pairs formed in the dye on irradiation by light and a transfer of the charge carriers to the systems which are in contact with each other.

3. In conformity with the last-named state of affairs, differences in the direction of the passage of electrons produced in the dye by light-irradiation when measurements were made in air and *in vacuo* (at 10^{-5} mm Hg) were also observed in the model silver halide/dye arrangements. The example given in Fig. 71 of the redrawn, recorded curves of an Ag/AgBr/Rhodamine B/Ag system reveals that whereas a photocurrent which corresponds to a transfer of electrons from the AgBr to the dye flows *in vacuo*, the polarity of the photovoltaic effect is observed to be reversed in air or oxygen, and this is indicative of a transfer of electrons from the dye to the semiconductor. In the example given, the photocurrent measured *in vacuo* at the commencement of the exposure

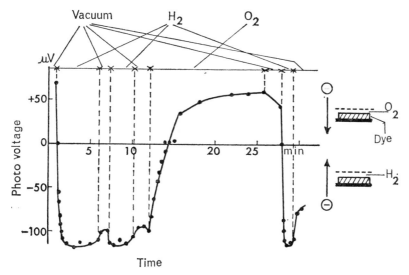

Fig. 70. Pinachrome (arrangement of the cells as in Fig. 68). The photovoltage as a function of the action of H_2, O_2 with time (100 Torr).

is moreover superimposed by a short (< 1 sec) reversed constituent of the current.

This result, which is in harmony with the above discussion, can be explained as follows:

a. The transfer of electrons from the semiconductor to the dye on irradiation in the absorption band of the dye and measured *in vacuo* is indicative of the action of a junction potential difference which is built up between the *n*-conducting dye and the silver halide in the unexposed state. The direction of motion of the electrons ascertained by measurement thereby conforms, with the desensitizing action of Rhodamine B which has been observed in a few cases.

b. The transfer of electrons in the reverse direction, namely from the dye to the AgBr, as measured in air, shows that the surface barrier layer which originates from the chemisorption of oxygen possibly has a direct influence on the model system which is an imitation of a spectrally sensitized photographic emulsion layer. These measurements prove that surface barrier layer effects of chemisorbed gases which may cause the transfer of electrons have to be reckoned with in dye-sensitized semiconductors (AgBr, etc.).

c. The short constituent of the current which occurred at the commencement of the exposure and which was originally attributed to the photovoltaic effect in the Ag/AgBr system which is of interest in connection with the mechanism of the primary photographic process,[343,344,378] can apparently be better explained by the action of residual chemisorbed oxygen which is not removed when the

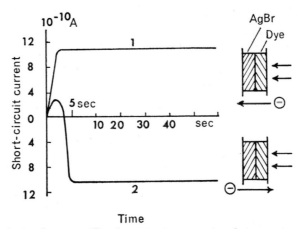

Fig. 71. AgBr/Rhodamine B-system. The short-circuit current in relation to time and to the action of air and a vacuum, respectively (10^{-5} Torr). (1) Measurement in air. (2) Measurement at 10^{-5} Torr. Radiation efficiency $300 \cdot 10^{13}$ quanta per $cm^2 \cdot sec$ ($\lambda = 6000$ Å). Recorded curve, redrawn.

gas is pumped off. This is because it is not until the electrostatic constituent of the potential which originated from the adsorbed "$O^{-(\sigma)}$-residues" has been removed that the potential gradient of the *pn*-junction which is also present *in vacuo* between the dye and the semiconductor can have any effect in bringing about the transfer of electrons corresponding to this field.

7.3 Concluding Remarks on Chapters VI and VII

None of the theories of the spectral sensitization of the photographic process and of the photoelectric effects in inorganic and organic photoconductors at present discussed is able to furnish an interpretation of the reactions which are characteristic of sensitization. Since the resonance theory cannot be brought into line with some of the essential conditions for sensitization to take place, an attempt was made to come nearer to solving the problem of the spectral sensitization of photochemical and photoelectric effects by further development of the electron transfer hypothesis.

This theory attributes the sensitization reaction to two effects: firstly to the formation and migration of the electronic charge carriers which are produced in the sensitizer by the absorption of radiation, and secondly to the formation of a junction potential difference in the zone which lies between the dye and the semiconductor and which separates the charge carriers formed by the action of light. Some of the results which were obtained with model arrangements of sensitized systems indicate that on the basis of this mechanism some of the conditions which are necessary for the sensitization reaction to take place might be satisfied in the case of sensitization with multimolecular and monomolecular layers of dye.

The said mechanism explains a number of experimental observations for which no interpretation was previously forthcoming, namely the spectral sensitization of p- and n-conductors by means of photographic desensitizers, sensitization of the photoconductivity of hole conductors, desensitization by thick layers of dye, the repeated reactivity of dye molecules, supersentization and other effects.

Any further experimental and theoretical investigations which are undertaken should aim at giving a quantitative coverage of the individual effects with the object amongst other things of acquiring a more profound knowledge of the mechanism of conduction in organic photoconductors, of the position of the energy levels in dyes and silver halides, of the electrical and optical behaviour of molecular layers of dye, and of the constitutional relationships which exist between the various effects.

References

[696] MEIER, H., in *Optische Anregung organischer Systeme*, (*Optical Excitation of Organic Systems*), edited by W. FOERST, p. 728, Verlag Chemie (Heidelberg) 1966.

[697] ALBRECHT, W., *Dissertation*, (*Thesis*), Universität Erlangen 1964.

[698] MEIER, H. and ALBRECHT, W., *Ber. Bunsenges. phys. Chem.*, **69**, 160 (1965).

[699] MEIER, H. and HAUS, A., *Angew. Chem.*, **72**, 631 (1960); *Z. Elektrochem.*, **64**, 1105 (1960).

[700] NELSON, R. C., *Symposium on Electrical Conductivity in Organic Solids*, p. 247, Interscience Pub. (New York) 1961.

[701] DOSSE, J., *Der Transistor* (*The Transistor*), R. Oldenbourg (München) 1959.

[702] HAMANN, C., *Phys. stat. sol.*, **4**, K 97 (1964).

[703] AIRAPETYANTS, A. V., VIOTENKO, R. M., DAVYDOV, B. E. and KRENTSEL', V. A., *Dokl. Akad. Nauk SSSR*, **148**, 605 (1963).

[704] FRANT, M. S. and EISS, R., USP 3,231,500 (1966).

[705] KATON, J. E., USP 3,222,469, Monsanto Co. (1966).

[706] VANNIKOV, A. V., *Dokl. Akad. Nauk SSSR*, **152**, 905 (1963).

[707] BEUN, M. and TUMMERS, L. J., *Philips tech. Rdsch.*, **25**, 69 (1963/64).

[708] HAAK, F. A. and NOLTA, J. P., *J. chem. phys.*, **38**, 2648 (1963).

[709] RUST, J. B., HAAK, F. A. and NOLTE, J. P., *WADD Techn. Rep. 60–111*, Hughes Res. Lab., Culver City, Calif., Oct. 1959.

[710] POHL, H. A., *Organic Semiconductors*, edited by J. J. BROPHY and J. W. BUTTREY, p. 139, Macmillan (New York) 1962.

[711] INOKUCHI, H., MARUYAMA, Y. and AKAMATU, H., *Bull. chem. Soc., Japan*, **34**, 1093 (1961).

[712] PUTZEIKO, E., *Dokl. Akad. Nauk SSSR*, **124**, 136 (1959).

[713] JORDAN, A. G., LADE, R. W. and SCHARFETTER, D. L., *Amer. J. Physics*, **31**, 490 (1963).
[714] CHAPIN, D. M., FULLER, C. S. and PEARSON, G. L., *J. appl. Phys.*, **25**, 676 (1954).
[715] HÄHNLEIN, A., *Nachrichtentech. Z.*, **9**, 145 (1956).
[716] KALMAN, J., *Electronics*, **32**, 5 and 59 (1959).
[717] WIESNER, R., *Halbleiterprobleme III (Semiconductor Problems) III*, edited by W. SCHOTTKY, p. 59, F. Vieweg & Sohn (Braunschweig) 1956.
[718] RUPPEL, W., *Phys. stat. sol.*, **5**, 657 (1964).
[719] DIEHL, H., *Phys. stat sol.*, **9**, 621 (1965).
[720] BECKER, M. and FAN, H. Y., *Phys. Rev.*, **78**, 301 (1950).
[721] CULVER, R. B. and SORENSEN, D. P., *J. chem. Phys.*, **42**, 2975 (1965).
[722] HEILAND, G., MOLLWO, E. and STÖCKMANN, F., *Solid State Phys.*, Vol. 8, edited by E. SEITZ and D. TURNBULL, p. 191, Academic Press (New York) 1959.
[723] DUDKOWSKI, S. J., KEPKA, A. G. and GROSSWEINER, L. I., *J. Phys. Chem. Solids*, **28**, 485 (1967).
[724] NEEDLER, W. C., *J. chem. Phys.*, **42**, 2972 (1965).
[725] AKIMOV, I. A. and TERENIN, A. N., *Dokl. Akad. Nauk SSSR*, **165**, 1332 (1965).
[726] USP 3,057,947, M. CALVIN and D. R. KEARNS (1962).
[727] USP 3,009,006, J. KOSTELEC (1961).
[728] USP 3,009,981, WILDI, B. S. and EPSTEIN, A. S. (1961).

[729] YOSHIDA, M., *Bull. Kobayasi Inst. phys. Res.*, **13**, 109 (1963).
[730] MEIER, H., *Chimia*, **18**, 179 (1964).
[731] GAFFRON, H. and WOHL, K., *Naturwissenschaften*, **24**, 81 and 103 (1936).
[732] BRODY, S. S., *Science*, **128**, 838 (1958).
[733] ROBINSON, G. W., *A. Rev. phys Chem.*, **15**, 311 (1964).
[734] CLAYTON, R. K., *Photophysiology*, Vol. I, edited by A. C. GIESE, p. 159, Academic Press (New York) 1964.
[735] TERENIN, A., PUTZEIKO, E., AKIMOV, I. and MESHKOV, A., *Dokl. Akad. Nauk SSSR*, **155**, 900 (1964).
[736] MATARE, H. F., *Z. Naturf.*, **10a**, 640 (1955).
[737] HARTEN, H. V. and SCHULTZ, W., *Halbleiterprobleme III, (Semiconductor problems) III*, edited by W. SCHOTTKY, p. 76, F. Vieweg & Sohn (Braunschweig) 1956.
[738] SHOCKLEY, W., *Electrons and Holes in Semiconductors*, p. 295, D. Van Nostrand (Princeton, New Jersey) 1950.
[739] KITTEL, C., *Solid State Physics*, p. 590, John Wiley & Sons, Inc. (New York) 1953.
[740] HEILMEIER, G. H. and HARRISON, S. E., *Phys. Rev.*, **132**, 2010 (1963).
[741] BUBE, R. H., *Photoconductivity in Solids*, p. 174, John Wiley & Sons, Inc. (New York) 1960.
[742] ENGELL, H. J., *Halbleiterprobleme I (Semiconductor Problems I)*, edited by W. SCHOTTKY, p. 249, F. Vieweg & Sohn (Braunschweig) 1954.
[743] PUTZEIKO, J. K. and TERENIN, A. N., *Dokl. Akad. Nauk SSSR*, **70**, 401 (1950).

INDEX

absorption coefficient 27
absorption of light by dyes
—, calculation 41
—, connection with constitution 40
—, influence of aggregation 34
—, theory 35
acceptor centres 99
acetonitrile 174
acid polymethine dyes 68
Acridine orange 41, 133
Acridine yellow 86, 133, 166
action spectrum
— of photoconductivity 122
— of photovoltaic effects 196, 204
activation energy
— of dark-conductivity 120
— of photoconductivity 122
adsorption
—, absorption change caused by 33
—, formation of H-bands 34
—, formation of J-bands 35
adsorption condition
— for sensitization 51, 52, 78, 83
— for desensitization 88
adsorption isotherms 81, 108
adsorption of dyes
— and conductivity of semi-conductors 202, 209
— and dye structure 83
—, mechanism 82, 108
—, number of dye molecules adsorbed 203, 205
AgBr
—, EPR after sensitization 110
—, fluorescence quenching by 109
—, forming of layers of 181
—, position of the conduction band 147, 175
—, sensitization 88
—, sensitized photoconductivity 98, 102, 104
— — and influence of gases 108
AgBr/dye system 182, 194, 200, 211
AgBr-photoelements 97, 98, 186
AgCl
—, sensitized photoconductivity 102
AgI
—, action spectrum of 204
—, fluorescence quenching of 109
—, p- and n-conduction of 99
—, sensitization 88
—, sensitized photoconductivity 98, 102, 104, 105
— — and influence of gases 108
AgI/dye systems 182, 200, 201
—, doping of 204

aggregation of dyes
—, influence on absorption 33
—, influence on relative quantum yield 28
—, relation to H-, J-bands 34, 82
—, structure 81
allopolar isomerism 73
Amethyst violet 41
ammoniacal emulsions 60
anils 168
anionic dyes 62, 68
anodic half-stage potential 39
— in dyes 173
—, in radical-ions 39
anthracene 111, 121, 165
—, mobility of charge carriers in 169
antiauxochrome groups 45
antisensitization 19, 50, 85, 89
ASA-system 17
Astraphloxin 42, 53, 54
asymmetrical dyes 45
—, cyanines 62
auxochrome 35, 45, 61, 78
azacyanine 43, 54, 88

barium arachidate 157, 205
barium stearate 147, 150
barrier layer 207
bathochromic effect
— and desensitization 52, 56
— definition 33
— on substitution 41, 44, 54
Becquerel-effect 98, 102, 103, 118, 126
benzimidazole 63
p-benzoquinone 108
benzoxazole 44
benzoxazoletrimethine cyanine 58
benzthia-undecarbocyanine 40
benzthiatrimethine cyanine 44
benzthiazole nucleus 44, 48, 64, 69
Betain cyanine 64
BF_3 133
bis-(1-ethylquinolyl-2)-trimethine cyanine; see Pinacyanol 64
bis-(1-ethylquinolyl-4)-trimethine cyanine 66
boiled emulsions 60
boundary layer 186
Brilliant green 131
p-bromanil 160
bromine 108, 146, 208
BSI-system 17

Callier coefficient 16
candela 14

215

Capriblue 166
carbocyanines
—, nomenclature 65
—, optimum quantity 74
Carotine 133
catalysis 117
cathodic half-stage potential 39
— connection with energy levels 87, 173
—, in radical-ions 39
— of thiacyanines 87
cationic cyanines
—, nomenclature 65
cationic dyes 62
CdO 102, 209
CdS
—/CdSe photocell 195
—, position of energy levels 175
— sensitization of 102, 103, 104, 106, 108
— —, quantum yield 107, 164
CdS/dye-systems 181, 182, 193, 194, 195, 199
— doping of 196
— stability 203
CdSe
—/CdS photocell 194
—/dye photocell 194
—, photoconductivity 181
Characteristic curve
—, construction 15, 16
—, influence of dye-sensitization on 28
—, monochromatic 17
charge carriers
— in sensitized photoconductivity 104
charge-transfer process 163
chemical potential 171, 190, 208
— and spectral sensitization 198, 202
— of sensitizing units 205
chemical sensitization 84
chemisorption
— of electron acceptors 208
— of hydrogen 210
— of oxygen 207
Child's law 124
chlor-indium-chlorophthalocyanine 134
—/α·Cu·phthalocyanin pn-system 186
o-chloranil 158, 160, 204
p-chloranil 109, 160, 204
chlorophyll 33, 104, 112, 133, 187
chromatographic purification 119, 131, 134
cis-trans isomers
—, sensitizing power of 73
colour couplers
—, influence on relative quantum yield 28
—, reaction with the sensitizer 52, 64
colour photographic materials
—, spectral sensitization of 5, 23, 52, 64, 70
condensor photoeffect, see Dember effect
— of thallium halides 208

conduction band, position in dyes 174, 175
Congo red 133
constitution
—, relation with sensitizing power 78
contact between semiconductors
— without light 188
— on irradiation 191
contact potential
—, change during illumination 102, 103
—, measurement 133, 170
— of dyes 135
crossed wedge method 16
crowded dye 58
crystal phosphors 130
crystal photoeffect
— in AgI 98
crystal-surface
—, influence on forming H-states 60
—, influence on forming J-states 35, 60
Crystal violet
—/amine systems 194
—, energy levels of 174
—, photoconductivity 126, 131, 133
—, sensitization of photoconductivity 108, 151
—, spectrum 173
CuI 108
Cu_2O 102
Cu-phenylacetylide 111
Cu-Phthalocyanine
— /β-chlor-indium-chlorphthalocyanine pn-system 186
—, photoconductivity 122, 128, 133
— /Rh-photocell 187
Cyanine 65
cyanines 62
—, D-bands of 34
—, half-stage potentials of 173
—, J-bands of 34
—, layer effect of 80
—, nomenclature 65
—, photoconductivity 128
—, synthesis of 62

dark conductivity
— and dye adsorption 209
— of silver halide 96, 97
D-bands 34
— photochemical activity of 82
degree of desensitization 26
degree of sensitization 26
Dember effect 102, 103, 108, 118
density
—, diffuse 16
—, measurement 15, 16
—, optical 15
densograph 16

desensitization
— and influence of oxygen 169
— and position of the excitation level 56, 87
— by multimolecular layers 80, 86, 181, 202, 209
—, connection with sensitization 86
—, connection with the basic material 88
—, definition 85
— of photoconductivity 108
—, *pn* theory of 202
—, temperature effect 88
desensitization of electro-photographic material 24
desensitization of silver halide
— by desensitizing dyes 19
— by radical-ions 40
— by sensitizing dyes 19, 23, 74
—, influence on the Schwarzschild effect 25
desensitizing dyes
—, bathochromic effect of 56, 166
—, chemical potential 198, 202
—, constitution 52–56, 86, 166
—, photoconductivity of 165
—, position of (1) excited level 87
—, redox potential of 87
—, sensitization of photoconductivity by 98, 106, 153
—, substitution effects 52, 54, 168
desoxynucleic acid 139
development and desensitization 87
diazocarbocyanine 55
dibromsuccinic acid 159, 160
dicarbocyanines 65
1,1′-diethyl-3,4-benz-2,2′-cyanine iodide 100
1,1′-diethyl-5,6-benz-2,2′-cyanine iodide 100
1,1′-diethyl-6-bromo-2,2′-cyanine iodide
—, triplet formation 84
1,1′-diethyl-2,2′-cyanine 34
—, aggregation of 34
—, antisensitization of 92
—, supersensitization of 89, 109
—, triplet formation 84
1,1′-diethyl-5,6,5′,6′-dibenzo-2,2′-cyanine 92
1,1′-diethyl-6-iodo-2,2′-cyanine iodide
—, triplet formation 84
N,N′-diethyl-2,2′-α-naphthothiazole carbocyanine 122
3,3′-diathylthiapentacarbocyanine 65
diffusion voltage 189, 190, 194, 207
2-(*p*-dimethylaminostyryl)-benzthiazole 90
2-(*p*-dimethylaminostyryl)-quinoline 90
m-dinitrobenzene 160
diphenylmethane dyes 62
α-diphenyl-β-picrylhydrazil 121, 122
DIN-unit 17
direction dependence of conductivity 118
donor-acceptor complexes 110

donor centres 99, 202
doping effect 159, 161
— and supersensitization 203
— of dye/semiconductor systems 196
drift mobility
— of holes in silver halide 96
— of photoelectrons in silver halide 96
dye/metal contact 187
dye/semiconductor photovoltaic effects 182
dyes
— chemical potential 198, 202
—, classification 169
— connection between conductivity and photographic properties 169
—, dark-conductivity 120
—, *n*- and *p*-type 133, 166, 205
—, photoconductivity 128

electrochemical half-stage potentials 173
electrochemical potential 172, 208
electron-acceptors 160
electron affinity
— definition 172
— of dyes 170, 174
electron gas model
— and conductivity 121
—, calculation of absorption bands 39, 45
—, theory of 35
electron microscope investigations
—, in silver halide grains 96
electron spin resonance, *see* EPR
—, and doping 164
—, and formation of the latent image 96
—, and sensitization 110
electronic theory of sensitization 151
electrons
—, mobility of in silver halides 96
electrophotographic materials
—, desensitization 24
—, measurement of spectral sensitivity distribution 22, 24
—, organic 110
—, removing dyes from 74
—, sensitization 74, 101
—, supersensitization 24
electrophotographic measurements 103
electrostatic potential, *see* macropotential
emulsions
—, high-speed 18
—, low-speed 18
—, sensitizing of 73
energy bands
— in organic semiconductors 137
energy gap 172
— and half-stage potentials 174
energy level diagram
— for *n*- and *p*-conducting dyes 136

217

energy in AgBr, ZnO   175
— of *pn*-junctions   188, 207
energy levels
— and redox potentials   87
— and sensitizing and desensitizing action   87
—, displacement on adsorption   83
—, position of   147, 151, 170, 174, 175
energy transfer
— and antisensitization   92
—, efficiency   50
— in spectral sensitization   82, 84
—, radionless   147, 149
Eosine   52, 104, 106, 111, 131, 133
— /ZnO system   193
EPR   110
equilibrium, condition for formation   191
Erythrosine   68, 104, 106
—, photoconductivity   133
—, repeated sensitizing power   107
—, work function of   135, 174
—, ZnO sensitization   108
3-ethyl-2-(*p*-dimethyl-aminostyryl) benzthiazole   109
Ethyl red   66
exchange membranes, photocells of   196
exciton hypothesis of conductivity   141
exposure E   14
exposure range   16
external photoeffect   134
eye sensitivity distribution   18, 23

Fermi energy   171
Fermi levels
—, determination   134
— in dyes   134
filters
— for colour correction   23
— for integral sensitometry   15
Flavine   133
Fluorescein   68, 104, 108, 111, 131, 133, 167
fluorescence
—, deactivation by   50
— of J-bands   34
— of adsorbed sensitizers   83
—, quenching of   19, 109, 147, 156
—, radiation transformation by   24, 145
—, sensitized   147
fluorescence of the 2nd kind   83
fog   16
fogging action of dyes   51, 78, 88
Freundlich adsorption isotherm   108
Fuchsine   104, 133

gain factor   126, 128, 165
gamma
— in unsensitized and sensitized emulsions   24
—, influence by J-aggregates   60

gamma (*contd.*)—
— monochromatic   24
—, relation to unexposed emulsion density Δ 29
gases
—, barrier layers with   207
—, influence on conduction in dyes   131
—, influence on spectral sensitization   108, 146, 209
gelatin
—, influence on dye adsorption   82
—, influence on J-state   35
—, photoconductivity of   111
general conditions of spectral sensitization   198
general structures   73
geometrical prerequisites for sensitization   57
germanium   102, 136
glow curves   130, 131
gradation   16
grain size distribution
—, influence on gamma   30
—, influence on monochromatic gamma   24
Grotthus–Draper law   33
Gurney–Mott theory
—, latent image formation   97
— of spectral sensitization   152, 200

H. and D. degrees   17
H-bands
—, formation on different crystal surfaces   60
—, in sensitized photoconductivity   108
—, photochemical activity   82
haematoporphyrine, sensitization with   204
Hall effect
— in dyes   133, 136, 160
— in silver halides   96
Hammett constants   48, 49
— and conductivity   168
—, definition   168
hemicyanines   62
hemioxonoles   70
heptamethine dyes   65
heterocyclic ring systems   62
HgI_2 102
HgO   102, 104
H_2O 108
Hofmann's violet   119, 133
hole traps   84, 110
holes
—, chemical potential of   190
—, definition   99
— in organic semiconductors   112
—, injection from electrodes   165
— mobility in silver-halides   96
holopolar cyanines   71
holopolar dye modification   73
hopping mechanism   137

H-state
—, relative quantum yield in 28
—, absorption 34
hydridization 38
hydrogen
— chemisorption of 210
hydrophilic groups
—, influence of 64
hydrophobization 64
hyperconjugating effect 44
hyperpanchromatic sensitization 23
hypsochromic effect
— definition 33
— on substitution 41, 44, 54

impulse field method 96, 105
infra-chromatic sensitization 23
infra-red sensitive photographic materials 23
infra-red sensitizing dyes
—, examples 40, 65
—, fogging action 51
—, influence on reciprocity law-failure 25
—, optimum quantity 74
—, photoconductivity of 123
—, sensitizing of photoconductivity 104, 107
infra-red spectroscopy 82, 88
intensity of illumination 14
interstitial silver ions 97, 153
intersystem-crossing
—, definition 50
—, quantum yield of triplet formation by 84
iodine 108
—, doping with 159, 197
ionization, energy of 172
Iris blue 52
Isocyanine 66
isopaque curves 25
Isoviolanthrone 121

J-aggregate
—, absorption of 60
—, influence by crystal surfaces 60
—, influence on the contrast 60
— relation with dye structure 59
—, structure 81
J-band
—, and supersensitization of the 1st kind 89
—, displacement on adsorption 83
—, in dye photoconductivity 122, 158
—, in sensitized photoconductivity 108
J-state
—, absorption behaviour 34
—, hole trapping of 110
—, influence of the silver halide substratum 35
—, relative quantum yield in 28, 35
—, temperature influence on 34

Kendall–Riester rule 52, 56
Krypotocyanine 65, 66
— sensitization of photoconductivity 104, 107

Langmuir adsorption isotherm 81, 108
latent image
—, external and internal 100
— mechanism of formation 96
lattice defects 84
layer effect 80, 157, 181
Lewis-acid 46
Lewis-base 46
lifetime of charge carriers
— in dyes 135, 165
— in silver halides 96
life time of excitation 155
long wavelength limit 173
lumen 14
luminescence, delayed 110
Luvican, see poly-N-vinyl-carbazole
lux 14

macropotential, see electrostatic potential 170, 189, 208
majority carriers 99
—, determination of 103, 133
— in dyes 133
— in ZnO, TlI 104
Malachite green
— adsorption on semiconductors 209
— /AgI photocell 194, 200
—, photoconductivity 126, 133
—, sensitization of photoconductivity 104, 106
— — and influence of gases 108, 208
—, spectrum of 139
—, work function of 135, 174
Malachite green/merocyanine photocell 181, 184, 193, 194
mean free path 136
mean value rule 47, 48
merocyanines 69
—, absorption displacement at adsorption 83
— /AgI system 201, 204
— /CdS system 199
—, D-bands of 34
—, desensitization properties 52
—, doping of 158, 161, 197, 204
—, energy levels of 174
—, formulae 105
—, p-conduction 131, 133
—, photoconductivity 106
—, pn-system 180, 182, 193
—, sensitization of photoconductivity 106
—, synthesis 70
—, work function of 135, 174
meropolar dye modification 73
meso-substitution 55, 64, 71

mesomeric effect (mesomerism) 46, 72
mesomeric state 73
mesomethylbenzthiatrimethine cyanine 55
Methylene blue 53, 86
— adsorption on semiconductors 209
—, energy levels of 174
—, photoconductivity 131, 133, 167
—, sensitization of photoconductivity 99, 107, 111, 151
Methylene green 133
Methylene red 53
Methyl violet 111
Mg-Phthalocyanine 133
— /tetramethyl-p-phenylendiamine photocell 194, 196
Mitchell's theory 153
Mg-Phthalocyanine 111
mobility
— and SCLC 125
— in anthracene 169
— in dyes 136, 138
—, trap influence on 124
2,2'-monomethine cyanine 66
  see Pinacyanol
2,4-monomethine cyanine 66
  see Ethyl red
4,4'-monomethine cyanine 66
  see Krypotocyanine
monomethine dye 65
monomolecular dye layers
— photoconductivity of 205
— sensitization with 205
M-state
—, relative quantum yield in 28

NaCl 102
naphthalene 121
naphthothiazole 63
n-conduction
— and Fermi level 135
—, definition 99
—, in dyes 131, 133, 168
—, in organic photoconductors 112
—, sensitization of 153, 198
negative photoeffect 97
neocyanine 59, 71
—, photoconductivity of 123
neutral dyes 62
neutrocyanines 69, 70
NH_3 108, 133
nitrile dyes 69, 70
nitrobenzimidazole 88
nitrochlorbenztriazole 88
nitrosobenzene 151
N,N'-ethyl-meso-ethyl-thiacarbocyanines 64
nn'-photoeffect 194, 196
NO 133

nomenclature of sensitization 23
— of polymethine dyes 65
nonamethine cyanines
—, optimum quantity 74
nonionic polymethine dyes 69
numeric specification of sensitization 25

o-formic ester 66
Ohm's law 123, 162
organic semiconductors
—, quantum yield 126
—, sensitization of 110
orthochromatic sensitization 23
orthochrome T
—, photoconductivity of 130, 131, 133
—, work function of 135
orthoesters for synthesis 66, 70
orthopanchromatic sensitization 23
Ostwald ripening 61
oxa (abbr.) 66
oxazole 63
oxygen
— and desensitization 169
— and photoconductivity 108
—, chemisorption of 207, 208
— —, and sensitization 209, 210
oxonoles 68

panchromatic sensitization 23
Parafuchsine 132
paramagnetic resonance absorption 110, 141
Pauli principle 37
PbO 102, 109
p-conduction
— and Fermi level 135
—, definition 99
— in dyes 131, 133, 168
— in organic photoconductors 112
—, sensitization of 153, 198
π-electrons 35, 138
pentamethine dyes 65
—, optimum quantity 74
—, synthesis 67
permeability coefficient 155
Perylene 163, 186
π-π*-excitation 120, 131, 152
Phenosafranine 79, 86, 88
—, energy levels of 174
—, gas influence 169
—, photoconductivity 122, 133
—, sensitization of photoconductivity 98, 106, 151
— /TlI photocell 193
—, work function of 135, 174
phenylthiazole 63
Phloxine BBN 133, 210
phosphinines 38, 68, 70

phosphorescence 110
photochemical bleaching-out of dyes 118, 126
photoconductivity
— of dyes 116, 120, 157, 205
— of silver halides 88, 96
photoconductivity, spectral sensitized
—, by desensitizers 88, 98, 208
— of inorganic semiconductors 101, 105
— of silver halides 98
—, quantum yield of 107
—, supersensitization of 89, 91, 100
photoelectrets 130
photoelectric activation energy 172, 174
photoflash experiments
—, at adsorbed dyes 84
—, mobility measurement 138
photometric system
—, physical 13
—, visual 14
photosynthetic unit 205
photovoltage of pn-junctions 183, 193
photovoltaic effect 102, 103
— and spectral sensitization 201
— in inorganic/organic systems 180
— mechanism 187, 191
phthalic anhydride 109
Phthalocyanine
—, dark-conductivity 121, 122
— doping of 160
—, energy levels of 174
—, external photoeffect 134
—, Hall effect 136
—, p-conduction
—, photoconductivity 118, 128
—, pn-system 180, 182, 194
—, rectifiers 186
—, sensitization by 104
—, structure 139
—, work function of 135
Pinachrom 133, 211
Pinacyanol 41, 57, 58, 64
—, dark-conductivity 121, 156, 202
—, energy levels of 174
—, external photoeffect 134
—, fogging action 51
—, J-aggregate 59
—, photoconductivity 122, 126, 129, 131, 133
—, sensitization of photoconductivity 98, 106, 111, 151
—, work function of 135, 174, 202
Pinacyanol blue 187
Pinaflavol 55, 62
—, ZnO sensitization 74
Pinakryptol green 88, 169
Pinaverdol 126
planarity condition 57

pn-junction
— in organic systems 186
— theory 187
pn-photoeffect 187
pn theory of spectral sensitization 198
polar solvents
—, influence of 73
polarographic cathodic half-stage potential 172
poly-acetylene 111
polyacrynitrile pn-junction 186
poly-Cu-phthalocyanine 196
polyethylene pn-junction 186
polymers as conducting materials 111, 121, 159
polymethin dyes 35
— as sensitizers 61
—, calculation of absorption bands 39, 62
—, nomenclature 64
—, number of π-electrons in 38
— of higher order 70
—, radicals of 39
—, synthesis 66
poly-N-vinyl-carbazole 111, 163
polythienylene 157, 205
print-out effect 96, 107
Pseudocyanine 66, 81, 139
Pyranthrone 121
Pyrazolenine 63
pyrazolin-5-one 68
pyridine 46, 63
Pyrolle 63
Pyronine G 166

Q-factor see Callier coefficient 16
2-quanta process resp. quantum summation 145 148
quantum yield
—, absolute 78
— of dye photoconductivity 125, 126, 157, 161, 205
—, relative (see this)
—, >1 164
quenching of fluorescence 19, 156, 109, 147
— by supersensitizers 91
— relation with antisensitization 92
2,2'-quinapolymethine cyanines 174
quinoline 44

radiation quantities 13, 14
radiationless energy transfer 147, 149
radical-ions of dyes
—, absorption of 39
—, half-stage potentials of 39
—, photographic behaviour 40
rate constants
— of fluorescence 155
— of tunnel transfer 154

221

reciprocity law-failure 148
—, influence of desensitization 25
—, influence of infra-red sensitizer 25
—, influence of wavelength on 25
rectifiers, organic 186, 196
red sensitivity of silver-halide 84, 146
redox potential
— of desensitizing dyes 87, 88
regeneration of the dye 79, 84, 107, 145, 179, 203, 209
relative quantum yield
—, definition 27
—, in spectral sensitization 28, 50, 79, 148
—, influence of additional substances on 28
—, influence of dye aggregation on 28
—, relation to degree of sensitization 28
removal of dyes from emulsions 64
resonance hypothesis 146, 201
rhodacyanines 70
Rhodamin B
— /AgI photocell 201
— /CdS photocell 195
—, energy levels of 174
— photoconductivity 131, 133, 167, 168
— sensitization of photoconductivity 104, 106, 111
—, spectrum of 140
Rhodamine B/silver halide-system 182, 194, 201
— and influence of gases 211
rhodamines
—, D-bands of 34
—, J-bands of 34
—, n-conduction of 131
—, pn-systems 180
rhodanine 68, 69, 71
Rose Bengals
—, photoconductivity 133
—, sensitization of organic photoconductors 111, 158
—, supersensitization of 159
—, ZnO sensitization 74, 109
Rubrocyanine 65
100 mμ rule 40

Safranine 86, 133, 166
sandwich type photocell 102, 118
saturation of dye photo-currents 125
Seebeck effect, see thermoelectric effect 133, 140
selena (abbr.) 66
Selenium
—, sensitization of 102, 150
sensitivity 16
—, integral S_i 16, 17, 18
—, spectral $S\lambda$ 16, 18
sensitization efficiency 50
— and adsorption of gases 108
— and crystal surface 60

sensitization efficiency (*contd.*)—
—, influence of additives 52
—, influence of coupler molecules 52
—, influence of the thickness of the dye layer 80, 157
— of a single dye molecule 207
sensitization of the 2nd order 28, 35
sensitization spectra
—, calculation 35
—, measurement 19
sensitized fluorescence 147
sensitized photographic layers, models of 201
sensitizing dyes
—, chemical potential of 198, 202
—, connection with desensitizers 52, 56, 86
—, destruction by photolytic bromine 78
— for photoconductivity 105
—, optimum quantity in practice 74, 86
— —, and constitution 74
—, relationship between constitution and 52–55, 78
—, solubility of 51
—, spatial structure 57, 101
—, supersensitizing of 91
sensitometer 15
sensitometry 13
—, integral 14
—, spectral 14, 15
Sheppard-Brooker planarity rule 58, 59
short-circuit current 183, 192
Silver halide/dye-systems 182, 193, 199, 201, 202
— measurements in air and *in vacuo* 211
silver halide photoconductivity
— and primary photographic process 96
—, antisensitization 93
—, n- and p-conduction 99, 105
—, spectral sensitization 88, 96, 98
—, supersensitization 88, 100
singlet excitation
—, deactivation 50, 83
—, lifetime 135
singlet-singlet energy transfer 151
SnO$_2$ 102
SO$_2$ 133
Sodium salicylate 24
solubility of dyes 51, 52
space charge limited currents (SCLC) 124
spatial structure effects of dyes 57
spectral sensitivity curves
—, construction 19, 20
spectral sensitivity distribution
— in dye-sensitized silver halide 23
— in non-sensitized silver halide 22
—, measurement of 19
spectral sensitization
— and EPR 110

spectral sensitization (*contd.*)—
— connection with conductivity of dyes 156
—, connection with desensitization 86, 151, 167
—, connection with the basic material 88
—, general conditions 198, 206
— mechanisms 145
—, model systems of 180, 212
— of organic photochemical reactions 151
—, *pn* theory of 198, 200, 205
—, reaction time 109
spectrum
— of Malachite green 139
— of Rhodamin B 140
speed index 17
sublimation of dyes 118, 131, 134
substitution
—and sensitizing properties 63, 78
— at the ring nitrogen atom 63, 64
—, desensitization by 52
— in the conjugated chain 41, 64
— in the nuclei of dyes 44, 48
Sulphur 102
supersensitization
— of photoconductivity 91, 100, 109, 117, 158, 163
— of the 1st kind 89
— of the 2nd kind 89, 100
supersensitization of silver-halide
—, connection with doping-effects 203
—, definition of 19, 85, 89
—, temperature influence 79
supersensitizers 58, 71, 89, 90
surface barrier layer 208, 210
surface type photocell 102, 118
synthesis of sensitizing dyes 33, 62, 66
Scheibe's rule 151
— in dye layers 175, 206
Scheiner degree 17
Schottky boundary layer 193
Schumann plates 24
Schwarzschild effect 25
stabilizer
—, effect on sensitization 83
stereoisomeric effect, *see* allopolar isomerism 73
Stern-Volmer-equation 19
streptocyanines 62, 67
styryl dyes 62
symmetrical cyanines 62

temperature influence on
— desensitization 79, 88
— dye photoconductivity 130, 157
— mobility 138
— sensitization 79, 149, 157, 204
— sensitized photoconductivity 100, 107, 157
Tetracyanoethylene 160

tetraiodophthalic anhydride, doping with 159, 197
1,1′,3,3′-tetramethyl-2,2′-cyanine 92, 194
Tetramethyl-*p*-phenylenediamin/Mg-phthalocyanine photo-element 194, 196
thermal work function 135, 170, 174
Thermoelectric effect 133, 140
thia (abbr.) 66
thiacyanines
—, photographic behaviour 87
—, polarographic half-stage potential of 87
Thiadiazoles 111
thiapolymethine cyanines 174
thiazines
—, D-bands of 34
thiazole 63
Thiazole purple 54
thiazoline 63
thickness effect of dye layers 80
Thioflavine 133
thiohydantoin 68, 71
Thionine 108
thiooxazolidinedienone 68
Thomson vibration method 103, 170
TlBr 102, 108, 109
TlCl 102
TlI 102, 104, 106, 107, 108, 109, 157
— /dye systems 194
— position of energy levels 175
trap capture cross section 125
trap density 125
trap distribution 125
traps
— and decrease of photo-current 129
— and SCLC 124
— and temperature effects 130
—, determination of the properties of 125
— in *pn*-systems 184
tricarbocyanines 65
trimethine dyes 65
—, synthesis 66
trinuclear cyanines 44
triphenylmethane dyes 62
—, D-bands 34
—, *n*-conduction of 131
—, photoconductivity 128
— /phthalein photoelements 196
triplet state
— and conduction 135, 137
— and trap action 131
— of J-aggregate 110
—, participation in the sensitization reaction 50, 84
—, quantum yield of formation 84
triplet-triplet absorption 84
triplet-triplet energy transfer 151

Trypaflavin   108
tunnel effect   137, 154

ultra-violet (< 2200 Å) recording materials   24
undecarbocyanine   65
unexposed emulsion density Δ
—, definition   29
—, relation to gamma   29

valency band, position of   172
Victoria blue   133
vinyl groups   40
Violanthrene   163, 187
voltage dependence of conductivity   124, 125

water-solubility of dyes   52
wedge spectrograms   22
wedges   15

Wien-Poole effect   123
work function
— thermal   135, 170
Wurster's blue   39

xerography   22
X-ray emulsions   88
X-ray investigation   81, 139
ZnO
—, adsorption of dyes on   108, 209
—, as electrophotographic material   24, 74
—, EPR after sensitization   110
—, position of energy levels   175
—, sensitization of   52, 74, 104, 107
—, supersensitization of   109, 158
—, work function of   193
ZnO/dye systems   193
zone refining   120